本教材由国家自然科学基金项目(编号:31470407、81503185),中央本级重大增减支项目(编号:2060302),浙江省重点研发项目(编号:2018C02030)资助出版

English Reading of Pharmaceutical Botany Resource
药用植物资源学英语阅读

主　编　史钰军　王慧中

编　者　史钰军　王慧中　钟　玮　卢江杰
　　　　冯尚国　沈晨佳　孟一君　王睿修

浙江工商大学出版社 | 杭州
ZHEJIANG GONGSHANG UNIVERSITY PRESS

图书在版编目(CIP)数据

药用植物资源学英语阅读 / 史钰军,王慧中主编.
— 杭州:浙江工商大学出版社,2019.1
ISBN 978-7-5178-3068-9

Ⅰ.①药… Ⅱ.①史…②王… Ⅲ.①药用植物－植物资源－英语－阅读教学 Ⅳ.①Q949.95

中国版本图书馆 CIP 数据核字(2018)第 284732 号

药用植物资源学英语阅读
YAOYONG ZHIWU ZIYUANXUE YINGYU YUEDU
史钰军 王慧中 主编

责任编辑	张莉娅
封面设计	林朦朦
责任印制	包建辉
出版发行	浙江工商大学出版社 (杭州市教工路 198 号 邮政编码 310012) (E-mail:zjgsupress@163.com) (网址:http://www.zjgsupress.com) 电话:0571-88904980,88831806(传真)
排　　版	杭州朝曦图文设计有限公司
印　　刷	浙江全能工艺美术印刷有限公司
开　　本	787mm×1092mm　1/16
印　　张	16.75
字　　数	402 千
版 印 次	2019 年 1 月第 1 版　2019 年 1 月第 1 次印刷
书　　号	ISBN 978-7-5178-3068-9
定　　价	58.00 元

版权所有　翻印必究　印装差错　负责调换

浙江工商大学出版社营销部邮购电话　0571-88904970

前　言

英语是高等院校许多专业重要的公共必修课,对学生适应市场经济中的对外交流以及未来发展起着重要作用,也是人文素质教育的内容之一。英语教学应根据其专业特点和未来需要,选择教学内容和改革教学方法,以提高学生的综合素质,增强其迎接各种挑战的能力。

《药用植物资源学英语阅读》是"高等学校专门用途英语(ESP)系列教材"之一,是英语教学后续课程所使用的教材,是根据新时期大学英语教学的发展和新形势下我国高等教育人才培养目标的要求而开发的。该系列教材旨在将大学英语教学与学生所学专业相结合,以提高大学生的英语能力和专业英语水平,使生物学等专业的学生毕业后可以直接运用英语从事本专业相关的工作,或者继续学习深造,为进行学术研究及参加学术活动打下坚实的基础。

本书由杭州师范大学生命与环境学院王慧中教授和浙江工商大学杭州商学院史钰军副教授拟定纲要,参编人员有浙江工商大学杭州商学院史钰军,浙江工商大学外国语学院钟玮,杭州师范大学王慧中、卢江杰、冯尚国、沈晨佳、孟一君、王睿修等。其中引言、第一章至第三章、第六章由史钰军编写;第七章至第九章由钟玮编写;第四章由王慧中、卢江杰、冯尚国编写,第五章由王慧中、沈晨佳、孟一君编写,第十章由王睿修编写。每章围绕一个主题,再细化分为若干小主题,每个小主题有两篇阅读文章,各主题内容相互衔接,前后呼应,文章皆取材于最新的、前沿的、权威的论文、报刊、网络媒体等。很好地完成本教材的学习后,生物学等专业学生使用英语进行专业学习和学术交流的能力会得到较全面的提升。

本书的出版得到了国家自然科学基金项目(编号:31470407、81503185),中央本级重大增减支项目(编号:2060302)和浙江省重点研发项目(编号:2018C02030)的资助,在此表示衷心感谢。由于编者水平有限,而且成稿仓促,其中错误和缺点在所难免,恳切希望广大读者和同仁不吝赐教。

Contents

Introduction

Passage 1　Utilization of Traditional Medicine in China ·················· 1
Passage 2　Introduction to Medicinal Plants (Extract) ·················· 6

Chapter One　Species of Medicinal Plants Resources

1.1　**Species of Medicinal Plants Resources in China** ·················· 10
　Passage 1 Traditional Herbs of China ·················· 10
　Passage 2 Medicinal Plants from China ·················· 13
1.2　**Species of Medicinal Plants Resources in Other Countries** ·················· 15
　Passage 1 Introduction of Medicinal Plants Species with the Most Traditional Usage in Alamut Region ·················· 15
　Passage 2 Medicinal Plants List Ten Common Species in North America ·················· 19

Chapter Two　Distribution of Medicinal Plants

2.1　**Distribution of Medicinal Plants Resources in the World** ·················· 23
　Passage 1 Predicting the Distributions of Egypt's Medicinal Plants and Their Potential Shifts Under Future Climate Change (Part 1) ·················· 23
　Passage 2 Predicting the Distributions of Egypt's Medicinal Plants and Their Potential Shifts Under Future Climate Change (Part 2) ·················· 29
2.2　**Natural Distribution of Medicinal Plants Resources in China** ·················· 31
　Passage 1 Colonization and Diversity of AM Fungi by Morphological Analysis on Medicinal Plants in Southeast China (Extract) ·················· 31
　Passage 2 The Ability of Nature Reserves to Conserve Medicinal Plant Resources: A Case Study in Northeast China (Extract) ·················· 35
2.3　**General Situation of Medicinal Plants Resources in Administrative Regions of China** ·················· 38
　Passage 1 Plant Resources of Sichuan (Part 1) ·················· 38

Passage 2 Plant Resources of Sichuan (Part 2) ……………………………… 42
2.4 **Regionalization of Chinese Medicinal Plants Resources** …………………… 47
Passage 1 Regionalization of Chinese Material Medical Quality Based on Maximum Entropy Model: A Case Study of Atractylodes Lancea (Part 1) ………………… 47
Passage 2 Regionalization of Chinese Material Medical Quality Based on Maximum Entropy Model: A Case Study of Atractylodes Lancea (Part 2) ………………… 50

Chapter Three Reserves of Medicinal Plants Resources

3.1 **Concept of Medicinal Plants Resources Reserves** ……………………………… 54
Passage 1 Conservation of Medicinal and Aromatic Plants in Brazil (Extract) …… 54
Passage 2 Conservation Issues ……………………………………………………… 59
3.2 **Reserves of Main Medicinal Plants Resources** ……………………………… 62
Passage 1 Notable Chinese Medicinal Plants ……………………………………… 62
Passage 2 Predicting the Global Potential Distribution of Four Endangered Panax Species in Middle-and Low-Latitude Regions of China by the Geographic Information System for Global Medicinal Plants (Extract) ………………………… 67

Chapter Four Resources Chemistry of Medicinal Plants

4.1 **An Overview** ……………………………………………………………………… 70
Passage 1 Chemical Markers for the Quality Control of Herbal Medicines: An Overview (Part 1) ……………………………………………………………… 70
Passage 2 Chemical Markers for the Quality Control of Herbal Medicines: An Overview (Part 2) ……………………………………………………………… 75
4.2 **Brief Introduction to Chemical Constituents of Medicinal Plants Resources** ……… 80
Passage 1 Evaluation of Chemical Constituents and Important Mechanism of Pharmacological Biology in Dendrobium Plants (Part 1) ……………………… 80
Passage 2 Evaluation of Chemical Constituents and Important Mechanism of Pharmacological Biology in Dendrobium Plants (Part 2) ……………………… 85
4.3 **Main Biosynthetic Pathway and Research Examples of Chemical Constituents of Medicinal Plants Resources** ………………………………………………………… 89
Passage 1 Exploring Drug Targets in Isoprenoid Biosynthetic Pathway for Plasmodium Falciparum (Extract) ………………………………………………… 89
Passage 2 Traditional Uses, Chemical Constituents, and Biological Activities of Bixa Orellana L.: A Review (Extract) ……………………………………………… 93

Chapter Five Evaluation of Medicinal Plant Resources

5.1 **Quality Evaluation of Medicinal Plant Resources** ·················· 99
   Passage 1 Species Specific DNA Sequences and Their Utilization in Identification of Viola Species and Authentication of "Banafsha" by Polymerase Chain Reaction (Extract) ·················· 99
   Passage 2 Quality Evaluation of Ayurvedic Crude Drug Daruharidra, Its Allied Species, and Commercial Samples from Herbal Drug Markets of India (Extract) ·················· 102
5.2 **Benefit Evaluation of Medicinal Plant Resources** ·················· 106
   Passage 1 Sustainable Utilization of Traditional Chinese Medicine Resources: Systematic Evaluation on Different Production Modes (Part 1) ·················· 106
   Passage 2 Sustainable Utilization of Traditional Chinese Medicine Resources: Systematic Evaluation on Different Production Modes (Part 2) ·················· 111
5.3 **Evaluation of Genetic Diversity of Medicinal Plants** ·················· 116
   Passage 1 Genetic Diversity in Populations of the Endangered Medicinal Plant Tetrastigma Hemsleyanum Revealed by ISSR and SRAP Markers: Implications for Conservation (Extract) ·················· 116
   Passage 2 Genetic Diversity Study of Some Medicinal Plant Accessions Belong to Apiaceae Family Based on Seed Storage Proteins Patterns ·················· 120

Chapter Six Conservation and Sustainable Utilization of Medicinal Plant Resources

6.1 **Conservation Status of Medicinal Plant Resources** ·················· 124
   Passage 1 Utilization and Conservation of Medicinal Plants in China (Extract) ·················· 124
   Passage 2 Role of Biotechnology for Protection of Endangered Medicinal Plants (Extract) ·················· 128
6.2 **Conservation Strategies of Medicinal Plant Resources** ·················· 134
   Passage 1 Evaluation of Medicinal Plant Resources and Strategies for Conservation (Extract) ·················· 134
   Passage 2 Medicinal Plant Conservation ·················· 138
6.3 **Main Ways and Methods of Protecting Medicinal Plant Resources** ·················· 141
   Passage 1 Ecological Protection of Medicinal Woody Plants (Abstract) ·················· 141
   Passage 2 Protecting Traditional Medicinal Knowledge in Zimbabwe (Extract) ·················· 143

6.4 **Protection of Rare and Endangered Medicinal Plant Resources** ········ 147
    Passage 1 Endangered Species ········ 147
    Passage 2 Strategies to Protect Rare Australian Plants ········ 152
6.5 **International Conventions, Policies and Regulations Concerning the Protection of Biological Resources** ········ 154
    Passage 1 Convention on Biological Diversity (Extract) ········ 154
    Passage 2 Protecting Biological Resources—Implementation of the Nagoya Protocol in Europe ········ 159

Chapter Seven Regeneration of Medicinal Plant Resources

7.1 **Natural Regeneration of Medicinal Plant Resources** ········ 164
    Passage 1 Natural Regeneration Status of the Endangered Medicinal Plant, Taxus Baccata Hook. F. syn. T. Wallichiana, in Northwest Himalaya (Part 1) ········ 164
    Passage 2 Natural Regeneration Status of the Endangered Medicinal Plant, Taxus Baccata Hook. F. syn. T. Wallichiana, in Northwest Himalaya (Part 2) ········ 167
7.2 **Artificial Regeneration of Medicinal Plant Resources** ········ 170
    Passage 1 *in vitro* Regeneration of Hemidesmus Indicus L. R. Br., an Important Endangered Medicinal Plant (Extract) ········ 170
    Passage 2 An Improved Protocol for *in vitro* Propagation of the Medicinal Plant Mimosa Pudica L. (Extract) ········ 173
7.3 **Organ Regeneration of Medicinal Plant Resources** ········ 177
    Passage 1 What Processes Contribute to Organ Regeneration? ········ 177
    Passage 2 Rapid *in vitro* Plant Regeneration from Leaf Explants of Launaea Sarmentosa (Willd.) Sch. Bip. ex Kuntze ········ 179

Chapter Eight Development of Medicinal Plant Resources

8.1 **Drug Development** ········ 183
    Passage 1 Developing Drugs from Traditional Medicinal Plants ········ 183
    Passage 2 Screening for Natural Inhibitors in Chinese Medicinal Plants Against Glycogen Synthase Kinase 3β (GSK-3β) (Extract) ········ 186
8.2 **Development of Natural Care Products** ········ 190
    Passage 1 Five Spices and Herbs That Really Work for Treating Depression ········ 190
    Passage 2 Healing with Herbs, Grass and Flowers (Extract) ········ 193

8.3 Development of Health Food 197
    Passage 1 Natural Antioxidants in Foods and Medicinal Plants: Extraction, Assessment and Resources (Part 1) 197
    Passage 2 Natural Antioxidants in Foods and Medicinal Plants: Extraction, Assessment and Resources (Part 2) 200

8.4 Development of Botanical Pesticides 204
    Passage 1 Impacts of Synthetic and Botanical Pesticides on Beneficial Insects (Extract) 204
    Passage 2 The Toxicity, Persistence and Mode of Actions of Selected Botanical Pesticides in Africa Against Insect Pests in Common Beans, P. Vulgaris: A Review (Extract) 209

8.5 Development of Herbal Feed Additive 213
    Passage 1 Chinese Herbs as Alternatives to Antibiotics in Feed for Swine and Poultry Production: Potential and Challenges in Application (Part 1) 213
    Passage 2 Chinese Herbs as Alternatives to Antibiotics in Feed for Swine and Poultry Production: Potential and Challenges in Application (Part 2) 216

8.6 Development of New Medicinal Plants Resources 220
    Passage 1 National Medicinal Plants Board for Development of Medicinal Plants Sector 220
    Passage 2 Development of Medicinal Plant Gardens in Aburi, Ghana 221

Chapter Nine Investigation and Study on Medicinal Plant Resources

9.1 Purpose and Significance of Investigation and Study on Medicinal Plant Resources 226
    Passage 1 MaxEnt Modeling for Predicting the Potential Distribution of Endangered Medicinal Plant (H. Riparia Lour) in Yunnan, China (Part 1) 226
    Passage 2 MaxEnt Modeling for Predicting the Potential Distribution of Endangered Medicinal Plant (H. Riparia Lour) in Yunnan, China (Part 2) 230

9.2 Main Contents of Medicinal Plant Resources Investigation 233
    Passage 1 Biochemical Investigation of the Plant Terminalia Arjuna (Part 1) 233
    Passage 2 Biochemical Investigation of the Plant Terminalia Arjuna (Part 2) 236

9.3 Application of Modern Information Technology in Investigation of Medicinal Plant Resources 242
    Passage 1 Systems Medicine: The Application of Systems Biology Approaches for Modern Medical Research and Drug Development (Extract) 242

Passage 2 A System-Level Investigation into the Mechanisms of Traditional Chinese Medicine: Compound Danshen Formula for Cardiovascular Disease Treatment (Extract) ·········· 247

Chapter Ten   Introduction to Medicinal Plant Species

Passage 1 Preventive Effects of Dendrobium Candidum Wall ex Lindl. on the Formation of Lung Metastases in BALB/c Mice Injected with 26-M3.1 Colon Carcinoma Cells (Extract) ·········· 252

Passage 2 Scutellaria ·········· 257

Introduction

Passage 1

Utilization of Traditional Medicine in China

Introduction

The science of Chinese materia medica is a summary of experience of the Chinese labouring people of many centuries in their struggle against diseases, which takes an important role in traditional Chinese medicine, and has made great contributions to the development of both Chinese and world medicine. It has been widely popular in China and elsewhere for thousands of years.

Since the founding of the People's Republic of China, the Chinese government has formulated some policy and adopted important measures to protect and promote the development of Chinese materia medica. Especially, in recent ten years, our government has attached great importance to the development of traditional Chinese materia medica, and pointed out that traditional Chinese herbal medicine and modern medicine are of the same importance. Indeed, the development of Chinese materia medica has been listed in the state constitution.

In 1986, the State Administration of traditional Chinese medicine was set up under the leadership of the State Council.

In 1988, the State Administration of traditional Chinese medicine and materia medica was set up, in order to strengthen the administration of both traditional Chinese medicine and herbal medicine. It undertakes the tasks of health care, education, scientific research, international exchanges and cooperations, as well as the plantation, processing, manufacture and selling of the medicinal herbs.

At present, there are 530,000 medical and technical personnel in traditional Chinese medicinal field; 480,000 technical personnel and workers engaged in manufacturing, and managing the medicinal herbs. There are more than 2,000 hospitals of traditional Chinese medicine, and 170,000 beds within the hospitals. Practitioners of Chinese medicine have the same right to make diagnosis and prescription, just like the doctors of modern medicine. Their work has a legal status. In the rural areas, there are 1,300,000 village doctors who have received medical training, and can prevent and treat diseases of the peasants with acupuncture and herbal medicine. A medical network has been formed in counties, towns and villages. In China, there are more than 160 scientific research institutions of traditional Chinese materia medica, forming a scientific research system. There are more than 2,000 factories for manufacturing medicinal herbs, producing more than 4,000 kinds of ready-made Chinese herbal medicine every year. Thirty thousand enterprises of Chinese herbal medicine exist at present, forming a network of selling in the whole country. The State has set up 28 colleges of traditional Chinese medicine. The average number of students in each college is about 2,000. More than 80,000 qualified personnel of traditional Chinese medicine have been trained since the founding of the People's Republic of China.

The development of traditional Chinese medicine and drugs has exploited the rich resources of medicine in China, improved the health care of the people in our country, and created considerable economic benefits to the society.

The Research and Exploitation of the Resources of Medicinal Herbs

China is very rich in medicinal herbs. According to a general survey in China in recent years, it was found that, there are more than 5,000 kinds of herbs after investigating the variety, ecological environment and reserves. The State offers a land of 5,000,000 ha for planting the medicinal herbs every year, giving an output of about 250,000,000 kg per year. At present, there are 10,000 farms producing medicinal herbs and nearly 10,000 technical administrative personnel. A lot of patients in China prefer to receive traditional Chinese medicine and herbal medicine treatment. The total sales of Chinese medicinal herbs and western drugs are almost the same.

Chinese people can choose either traditional Chinese medicine, or modern medicine in preventing and treating diseases. This is a good advantage in improving the health care of the people in our country. In addition, as a newly developed industry, the production of herbs can create considerable economic benefits, and produce drugs for export. Up to now, China has established trade relations with more than 100 countries and regions. The average export exceeds 300,000,000 U. S. dollars every year.

The exploitation of traditional Chinese medicine resources extends beyond the development of the economy, and includes the advancement of medical science, and promotion of the people's health. For thousands of years, practice has shown that, the correct application of natural medicines from plants, animals and minerals does not only cure many commonly and frequently encountered diseases, but can also be effective on some very complicated, and serious diseases. For example, in the treatment of malignant tumours, cardiovascular diseases, and cerebrovascular hepatitis, traditional Chinese medicine is superior to western medicine in some aspects. traditional Chinese medicine also has better curative effects in treating diseases related to gynaecology, neurology, ophthalmology, dermatology and gastrointestinal disorders. Regarding diseases which have no effective treatment today, e. g. viral infections, immunopathy and functional diseases, traditional Chinese medicine often works. Moreover, traditional Chinese medicine is famous for its simple treating methods with less toxicity and side effects. Today, it is well established that drug-induced diseases do occur. Such cases are rare in traditional Chinese medicine, making it superior to orthodox western medicine.

In recent years, important progress has been made in the research of Chinese materia medica. This has been accomplished due to the application of new technologies in herbal research. For the study of effectively active elements of Chinese materia medica, according to imperfect statistics, more than 150 commonly used monomers of Chinese materia medica have been thoroughly studied in chemical and pharmacological aspects, etc., with modern methods. Over 500 active monomers have been separated from Chinese materia medica. It is found that there exists a batch of highly active elements. For instance, Qinghaosu is extracted from Qinghao (Herba Artemisiae Chinghao) with anti-malarial effect. This is an important breakthrough in the history of anti-malarial medicine after the discovery of quinine. The research on active elements of Chinese materia medica has demonstrated both the material base of efficacy of Chinese materia medica, and has discovered its new efficacy and usefulness. Thus the research on the making of new medicines in China has been pushed forward.

Centering on commonly-seen, frequently-encountered, and complicated and difficult diseases in China, research on complex prescription of Chinese materia medica (mutual combination of a group of medicines) has been launched. Its clinical results have been proved with experimental medicinal efficacy. The complex prescription of Chinese materia medica is one of the characteristics of prescribing traditional Chinese medicine. Its scientific essence is based on systematic regulation. Through scientific grouping and combining, it has such advantages as organic regulation and activation of human recovery ability, comparing with monomer medicine. The application of isotope, electronic mirror, biochemistry, immunology, technology of tissue culture, etc. to it, has promoted further explanation of its function and theory. By taking a medicine apart to analyse, it is found that in some complex prescriptions, among different flavor of herbs, there is function of cooperation and resistance to some degree.

Owing to various kinds of herbal medicine in China, wide-scale production in many places, a long history of application and many different habits, it has been reported that there has existed some confusion on the history of Chinese materia medica. Plants with the same name have been referred to as different herbs, while the same herbs have been assigned different names. In 1985 the State issued "Administrative Law of Medicines" and "Administrative Measures of Examination and Approval of New Medicines" to strengthen the administrative work of Chinese materia medica. This is a new requirement for management and research of Chinese materia medica. Meanwhile, the appraisal and research of Chinese materia medica have also been deepened, and it widened. In the past, the main ways of appraisal were based on properties of herbs and experience. At present, modern technology has been adopted, and it includes microscope appraisal, thin layer chromatogram, thin colorimeter, thin luminosity chromatogram, the combination of the chromatography and gas chromatography-mass spectrometer, and the highly effective liquid chromatography. These methods have solved the quality appraisal problem, ensuring the correctness and efficacy in clinical application of the medicines.

With the gradual change of the focal point of health work from treatment to prevention, the scope for exploiting and broadening the use of Chinese materia medica has also constantly become wide, and increasingly more penetrating to many aspects of daily life, such as health care food, health care drinks, cosmetology, cosmetics, medicinal food, natural pigment, natural sweet and bitter pharmaceutical preparations. On the whole, with the constantly rising of living standard of the people, the trend of thought "That Mankind Wants to Go Back to Nature" becomes more realised. For the development of these fields, Chinese materia medica has very great potency.

Some aspects need more attention in order to exploit herbal medicinal resources, and at the same time, to protect and put the resources to rational use. The State Council of the Chinese government has issued some regulations to manage and protect national wild medicinal resources. Chinese materia medica protection stations have been set up in the main production areas of certain medicinal materials that are managed by law. Meanwhile, numerous scientific research personnel are using scientific ways to constantly raise the quantity and quality of the specific resources.

However, the task confronting the medical workers, e. g. in controlling the population, and making it fit the rate of the development of the resources, the economy and the society is very hard. China needs to continuously improve the quality of population, and raise its health level; to prevent diseases, especially those serious ones such as cardiovascular, cerebro vascular, malignant tumour, respiratory, infectious, parasitic, and endemic diseases. Attention should also be paid to senile health, as one of the important social problems, especially the aging trend of the population structure. These concern not only the national economy, and the people's livelihood, but also reproduction of the coming generations, and the flourishing of the country. We believe that traditional Chinese medicine will definitely make greater contributions to the health of the world peoples.

The policy of the country to open up to the outside world has provided a wide scope for international exchange and cooperation in the field of traditional Chinese medicine. China is a developing country. It has traditional friendship, and friendly cooperation with many other developing countries. We would like to carry out various forms of cooperation and exchange with those countries or scholars who are interested in traditional medicinal plants, and make joint effects to promote the mankind in developing and putting traditional medicinal plants to rational use.

http://www.nzdl.org/gsdlmod? e=d-00000-00-off-0cdl-00-0-0-10-0-0-0direct-10-4-0-1l-11-kn-50-20-preferences-00-0-1-00-0-4-0-0-11-10-0utfZz-8-00&a=d&c=cdl&cl=CL4.119&d=HASHa9287526d39203650f9874.8.4

Vocabulary

Chinese materia medica　中药学；本草书
the state constitution　国家宪法
the State Administration of traditional Chinese medicine　国家中医药管理局
the State Council　国务院
technical personnel　技术人员
ecological environment　生态环境
malignant 英 [məˈlɪgnənt] 美 [məˈlɪgnənt] *adj.* 恶性的；致命的
cerebrovascular hepatitis　脑血管性肝炎
gynaecology 英 [ˌgaɪnəˈkɒlədʒi] 美 [ˌgaɪnəˈkɑːlədʒi] *n.* 妇科；妇科学；妇科医学
neurology 英 [njʊəˈrɒlədʒi] 美 [nʊˈrɑːlədʒi] *n.* [医]神经病学；[医]神经病学家
ophthalmolog 英 [ˌɒfθælˈmɒlədʒi] 美 [ˌɑːfθælˈmɑːlədʒi] *n.* 眼科学
dermatology 英 [ˌdɜːməˈtɒlədʒi] 美 [ˌdɜːrməˈtɑːlədʒi] *n.* 皮肤科；皮肤病(学)
gastrointestinal disorders　消化道疾病；胃肠功能紊乱
monomer 英 [ˈmɒnəmə] 美 [ˈmɒnəmə] *n.* 单体
isotope 英 [ˈaɪsətəʊp] 美 [ˈaɪsətoʊp] *n.* [化]同位素
potency 英 [ˈpəʊtnsi] 美 [ˈpoʊtnsi] *n.* 效力；潜能；权势；(男人的)性交能力

Useful Expressions

1. diagnosis 英 [ˌdaɪəgˈnəʊsɪs] 美 [ˌdaɪəgˈnoʊsɪs] *n.* 诊断；诊断结论；判断；结论
I need to have a second test to confirm the diagnosis.
我需要再进行一次检查以确诊。
2. prescription 英 [prɪˈskrɪpʃn] 美 [prɪˈskrɪpʃən] *n.* 处方药；[医]药方，处方
You will have to take your prescription to a chemist.
你得拿着处方去找药剂师。
3. exceed 英 [ɪkˈsiːd] 美 [ɪkˈsid] *vt.* 超过；超越；胜过；越过……的界限
Its research budget exceeds $700 million a year.
其研究预算每年超过7亿美元。

Questions

1. What is Chinese materia medica?
2. How many people are engaged in traditional Chinese medicinal field now?
3. How many kinds of herbs are there in China?

4. Why is Chinese medicine famous?

5. Why did the state issue "Administrative Law of Medicines" and "Administrative Measures of Examination and Approval of New Medicines"?

Passage 2

Introduction to Medicinal Plants (Extract)

Among 250,000 higher plant species on earth, more than 80,000 species are reported to have at least some medicinal value and around 5,000 species have specific therapeutic value.

Herbs are staging a comeback, and herbal "renaissance" is happening all over the globe. The herbal products today symbolize safety compare to the synthetics that are considered as unsafe to human and environment. Even though herbs had been priced for their medicinal, flavoring and aromatic qualities for centuries, the synthetic products of the modern age surpassed their importance, for a while. However, the blind dependence on synthetics is over and people are returning to the herbals with hope of safety and security. Over three-quarters of the world population relies mainly on plants and plant extracts for health care. More than 30% of the entire plant species were used for medicinal purposes.

Herbals in World Market

It is estimated that world market for plant derived drugs may account for about Rs. 2,00,000 crores. Presently, Indian contribution is less than Rs. 2,000 crores. The annual production of medicinal and aromatic plant's raw material is worth about Rs. 200 crores. This is likely to reach US $5 trillion by 2050. It has been estimated that in developed countries such as the United States, plant drugs constitute as much as 25% of the total drugs, while in fast developing countries such as China and India, the contribution is as much as 80%. Thus, the economic importance of medicinal plants is much more to countries such as India than to rest of the world.

Biodiversity of Herbals in India

India is one of the world's 12 biodiversity centers with the presence of over 45,000 different plant species. India's diversity is UN compared due to the presence of 16 different agro-climatic zones, 10 vegetation zones, 25 biotic provinces and 426 biomes (habitats of specific species). Among these, about 15,000—20,000 plants have good medicinal value. However, only 7,000—7,500 species are used for their medicinal value by traditional communities.

In India, drugs of plant origin have been used in traditional systems of medicines such as Unani and Ayurveda since ancient times. The Ayurveda system of medicine uses about

700 species, Unani 700, Siddha 600, Amchi 600 and modern medicine around 30 species. About 8,000 herbal remedies have been included in Ayurveda. The Rig-Veda (5,000 BC) has recorded 67 medicinal plants, Yajurveda 81 species, Atharvaveda (4,500—2,500 BC) 290 species, Charak Samhita (700 BC) and Sushrut Samhita (200 BC) had described properties and uses of 1,100 and 1,270 species respectively, in compounding of drugs and these are still used in the classical formulations, in the Ayurvedic system of medicine.

Sources of Medicinal Drugs

The drugs are derived either from the whole plant or from different organs, like leaves, stem, bark, root, flower, seed, etc. Some drugs are prepared from excretory plant product such as gum, resins and latex. Plants, especially used in Ayurveda can provide biologically active molecules and lead structures for the development of modified derivatives with enhanced activity and/or reduced toxicity. Some important chemical intermediates needed for manufacturing the modern drugs are also obtained from plants (e. g. $β^2$-ionone). The forest in India is the principal (diosgenin, solasodine) repository of a large number of medicinal and aromatic plants, which are largely collected as raw materials for manufacture of drugs and perfumery products. The small fraction of flowering plants that have so far been investigated have yielded about 120 therapeutic agents of known structure from about 90 species of plants. Some of the useful plant drugs include vinblastine, vincristine, taxol, podophyllotoxin, camptothecin, digitoxigenin, gitoxigenin, digoxigenin, tubocurarine, morphine, codeine, aspirin, atropine, pilocarpine, capscicine, allicin, curcumin, artemisinin and ephedrine among others.

History of Herbal Medicine

Ayurveda, Siddha, Unani and Folk (tribal) medicines are the major systems of indigenous medicines. Among these systems, Ayurveda is most developed and widely practiced in India. Ayurveda dating back to 1,500—800 BC has been an integral part of Indian culture. The term comes from the Sanskrit root Au (life) and Veda (knowledge). As the name implies it is not only the science of treatment of the ill but also covers the whole gamut of happy human life involving the physical, metaphysical and spiritual aspects. Ayurveda is gaining prominence as the natural system of health care all over the world. Today this system of medicine is being practiced in countries like Nepal, Bhutan, Sri Lanka, Bangladesh and Pakistan, while the traditional system of medicine in the other countries like Mongolia and Thailand appear to be derived from Ayurveda. Phytomedicines are also being used increasingly in Western Europe. Recently the US Government has established the Office of Alternative Medicine at the National Institute of Health at Bethesda and its support to alternative medicine includes basic and applied research in traditional systems of medicines such as Chinese, Ayurvedic.

Disadvantages

A major lacuna in Ayurveda is the lack of drug standardization, information and quality control. Most of the Ayurvedic medicines are in the form of crude extracts which are a mixture of several ingredients and the active principles when isolated individually fail to give desired activity. This implies that the activity of the extract is the synergistic effect of its various components. About 121 (45 tropical and 76 subtropical) major plant drugs have been identified for which no synthetic one is currently available.

The scientific study of traditional medicines, derivation of drugs through bioprospecting and systematic conservation of the concerned medicinal plants is of great importance.

Unfortunately, much of the ancient knowledge and many valuable plants are being lost at an alarming rate. *Red Data Book of India* has 427 entries of endangered species of which 28 are considered extinct, 124 endangered, 81 vulnerable, 100 rare and 34 insufficiently known species. There are basically two scientific techniques of conservation of genetic diversity of these plants. They are the in-situ and ex-situ method of conservation.

In-Situ Conservation of Medicinal Plants

It is only in nature that plant diversity at the genetic, species and eco-system level can be conserved on a long-term basis.

It is necessary to conserve in distinct, representative bio-geographic zones inter and intra specific genetic variation.

http://smallfarms.cornell.edu/2011/01/09/traditional-chinese-medicine-in-north-america-opportunities-for-small-farms/

Vocabulary

synthetic 英 [sɪnˈθetɪk] 美 [sɪnˈθɛtɪk] n. 合成物；合成纤维；合成剂
biodiversity 英 [ˌbaɪəʊdaɪˈvɜːsəti] 美 [ˌbaɪoʊdaɪˈvɜːrsəti] n. 生物多类状态，生物多样性；
Ayurveda 英 [ˈaɪjʊəˌveɪdə] 美 [ˌɑjʊəˈveɪdə；ˌɑjʊəˈvidə] n. 阿育吠陀
resin 英 [ˈrezɪn] 美 [ˈrezn] n. 树脂；合成树脂；松香
lacuna 英 [ləˈkjuːnə] 美 [ləˈkjunə] n. 腔隙；缺陷；空白；空隙
perfumery 英 [pəˈfjuːməri] 美 [pərˈfjuːməri] n. 香料店
inter and intra specific genetic variation 种间和种内遗传变异

Useful Expressions

1. prominence 英 [ˈprɒmɪnəns] 美 [ˈprɑːmɪnəns] n. 突出；声望；卓越
He came to prominence during the World Cup in Italy.
他在意大利的世界杯赛中声名鹊起。

2. synergistic 英 [ˌsɪnəˈdʒɪstɪk] 美 [ˌsɪnəˈdʒɪstɪk] adj. 增效的；协作的；互相作用（促进）的
The combination of HMHEC and cationic polymers surprising results in a synergistic effect.

HMHEC 和阳离子聚合物的结合在协同效应方面产生了令人惊奇的效果。

Questions

1. How many species of plants are said to have medicinal value?
2. Where can drugs be got from plants?
3. What disadvantages do traditional medicines have?

Chapter One

Species of Medicinal Plants Resources

1.1 Species of Medicinal Plants Resources in China

Passage 1

Traditional Herbs of China

With an estimated 30,000 flowering plant species, China possesses a diverse and rich flora to form the basis of traditional Chinese medicine. At least 5,000 of those plant species are used as medicinal plants in China, and many of those are used as food. The line between medicinal food and medicinal plants is blurry in China, as it is in all herbal traditions.

History

In China, traditional medicine is an integral part of the culture going back at least 5,000 years. In the 1970s interest in herbal medicine began to reemerge and with it many Chinese herbal remedies began appearing on health food store shelves.

The acceptance of herbal medicine in the west owes a debt to the Asian cultures. Many Chinese medicinal plants are well known, most notably Panax ginseng and Angelica sinensis (*dan-gui*), which have counterpart species here in North America, American ginseng, Panax quinquefolium and angelica Angelica archangelica. Indeed American gardens are a rich source of Chinese medicinal plants, day lilies, chrysanthemums and peonies to name a few, however, these plants are valued here only as ornamentals.

Principles

traditional Chinese medicine is a combination of energy theories, acupuncture, and herbal prescriptions. The Chinese system distinguishes between plants used as "drugs" and "herbs". While about 500 plants are used as official drugs, there are roughly 4,500 plants used in folk medicine by the people of the countryside. Root drugs are considered most important, followed by seeds and fruits, with leaf drugs considered least important.

Properties

The four properties are cold, hot, warm, and cool. The general principle of treatment is to treat heat-syndrome conditions like fever with cold-natured drugs, and patients with cold-syndrome diseases with warm drugs.

Flavors

The five flavors of traditional Chinese medicine are sour, bitter, sweet, hot, and salty. Acrid (pungent or hot) has the function of expelling cold. Sour has an astringent function. Sweet has the ability to alleviate pain. Bitter drugs help to harden and dry tissue. Salty drugs are able to soften hard lumps, such as tumors. Some herb drugs have more than one flavor; the first or primary one is used to determine its uses.

Trends of Herb Functions

There are four trends of drug functions: lifting, lowering, floating, and sinking. Diseases are also classified by direction, for instance, a cough rises from bottom to top. In traditional Chinese medicine the practice is to use a herb that moves in the opposite direction as the condition.

Attributive Channels

These channels are the particular meridian, organ or group of organs in the body that the herb has a major effect on.

In western herbalism, we often use "simples", that is a single herb, to treat a condition. Chinese medicinal prescriptions often use 3 or more herbs and can consist of up to 50 or more herbs. These prescriptions are formulated using a complicated system of the monarch, or main herb, assistants and guide drugs. Because of this complexity, it is not possible for the home practitioner to get the full benefit of Chinese herbal drugs without guidance. In practice I recommend that those who wish to examine this ancient system of medicine do so under the guidance of a Chinese herbal practitioner. This will allow you to get the full benefit in person diagnosis, and a comprehensive treatment program that combines acupuncture treatment with herbal prescriptions.

https://www.anniesremedy.com/chinese_herbs.php

Vocabulary

flora 英 ['flɔːrə] 美 ['flɔːrə] *n.* 植物区系；(某地区或某时期的) 植物群
blurry 英 ['blɜːri] 美 ['blɜːɪ] *adj.* 模糊的
integral 英 ['ɪntɪɡrəl] 美 ['ɪntɪɡrəl, ɪnˈtɛɡrəl] *adj.* 完整的；不可或缺的
integration 英 [ˌɪntɪˈɡreɪʃn] 美 [ˌɪntɪˈɡreʃən] *n.* 整合；一体化；结合
lily 英 ['lɪli] 美 ['lɪli] *n.* 百合花
chrysanthemum 英 [krɪˈsænθəməm] 美 [krɪˈsænθəməm, -ˈzæn-] *n.* 菊花
peony 英 ['piːəni] 美 ['piəni] *n.* 牡丹；芍药
property 英 ['prɒpəti] 美 ['prɑːpərti] *n.* 特性；属性
syndrome 英 ['sɪndrəʊm] 美 ['sɪndroʊm] *n.* 综合征；综合症状
flavor 英 ['fleɪvə] 美 ['flevə(r)] *n.* 味
acrid 英 ['ækrɪd] 美 ['ækrɪd] *adj.* 辛辣的；刺鼻的
pungent 英 ['pʌndʒənt] 美 ['pʌndʒənt] *adj.* 辛辣的；刺激性的

astringent 英 [ə'strɪndʒənt] 美 [ə'strɪndʒənt] n. [药] 收敛剂；止血药
attributive 英 [ə'trɪbjətɪv] 美 [ə'trɪbjətɪv] adj. 属性的
meridian 英 [mə'rɪdiən] 美 [mə'rɪdiən] n. 顶点；经络；脉
herbalism 英 ['hɜːblɪzəm] 美 ['ɜːrblɪzəm] n. 草药医术学

Useful Expressions

1. strive 英 [straɪv] 美 [straɪv] vi. 努力奋斗；力求
He strives hard to keep himself very fit.
他努力地保持身体健康。
2. validate 英 ['vælɪdeɪt] 美 ['vælɪˌdet] vt. 证实；使生效；批准；确认
This discovery seems to validate the claims of popular astrology.
这个发现似乎能印证流行占星术的一些说法。
3. estimate 英 ['estɪmət] 美 ['ɛstəˌmet] vt. 估计；评价
Try to estimate how many steps it will take to get to a close object.
尽力估算一下达到一个近距离目标需要多少步。
4. alleviate 英 [ə'liːvieɪt] 美 [ə'liviˌet] vt. 减轻；缓和
Nowadays, a great deal can be done to alleviate back pain.
如今，减轻背部疼痛有许多方法。
5. expel 英 [ɪk'spel] 美 [ɪk'spɛl] vt. 驱逐；把……除名；排出（气体等）
An American academic was expelled from the country yesterday.
昨天一位美国学者被逐出该国。
6. at odds 争执；不一致
It's possible to be bedfellows with someone on one issue, and at odds with him/her on another.
可能在一个问题上与某人达成一致，在另一个问题上又意见相左。

Questions

1. How many species are used as medicinal plants in China?
2. What cause the successful integration of traditional Chinese medicine and modern western style medicine in China?
3. What is included in traditional Chinese medicine?
4. What is the treatment principle of traditional Chinese medicine?
5. What is the difference between western herbalism and traditional Chinese medicine?

Passage 2

Medicinal Plants from China

Camptotheca acuminata, a deciduous tree native to southern China, contains camptothecin (CPT). In 1996, the Food and Drug Administration (FDA) approved two CPT derivatives for treating ovarian and colorectal cancer. In addition, other derivatives of CPT are being tested in clinical trials against other types of cancer in the United States. Manufacturing of the two anti-cancer drugs continues to rely on extraction from plant materials harvested mainly from naturally grown trees. There is no known plantation production in an agricultural setting in the United States or in China to supply the plant materials.

The medicinal plant research program at the Louisiana State University (LSU) Agricultural Center initiated a study using Camptotheca acuminata to develop a production system. Cultivation of Camptotheca acuminata not only presents cropping opportunities but also allows the input of effective production management that offers superior quality raw materials to the now available natural sources. Since 1993, plantations of Camptotheca acuminata have been grown in southern Louisiana, and extensive growth studies have been done to find cultural practices to enhance the levels of CPT. The plantations, however, were established with propagules from a single tree that is perhaps the offspring of an earlier introduction program of the US Department of Agriculture. Therefore, the genetic base is narrow. Camptotheca acuminata, however, has a broad range of natural distribution in China, covering almost all areas south of the Yangtze River.

The vast distribution area led the researchers to believe that natural variations in terms of growth and camptothecin concentrations exist. Consequently, cooperative research between the LSU Agricultural Center and Zhejiang Forestry College (now Zhejiang A&F University) was conducted to find these variations and to lay the foundation for clonal line development. Once the clonal lines are developed, specific cultivation practices using these clonal lines can be developed for high quality raw plant material production. This provenance study, which means an investigation of variations associated with geographical source of Camptotheca acuminata, is the first step in the effort to cultivate Camptotheca acuminata in both Louisiana and Zhejiang of China.

Within its natural distribution, 18 seed sources were collected from 10 provinces south of the Yangtze River. Trees 20 to 29 years old bearing seeds were selected from a large local area in November 2009. In Huzhou City, Zhejiang, China, nursery beds were prepared for the study.

Results showed significant variations among the 18 seed sources in growth rate and leaf CPT concentrations. Since leaves are the target plant materials, based on the findings of this study, the top five seed sources were identified. These seed sources warrant further growth studies to determine the optimal growth conditions for accumulating CPT.

http://www.lsuagcenter.com/portals/communications/publications/agmag/archive/1999/summer/medicinal-plants-from-china

Vocabulary

Camptotheca acuminate 喜树
deciduous 英 [dɪˈsɪdʒuəs] 美 [dɪˈsɪdʒuəs] adj. （指树木）每年落叶的
camptothecin 英 [ˌkæmptəʊˈθiːsɪn] 美 [ˌkæmptoʊˈθiːsɪn] n. 喜树碱（可用以治疗癌症）
ovarian 英 [əʊˈveərɪən] 美 [oˈvɛrɪən] adj. 卵巢的；子房的
colorectal 英 [ˌkəʊləˈrektəl] 美 [ˌkoʊləˈrektəl] adj. 结肠直肠的
crop 英 [krɒp] 美 [krɑːp] vt. 种植；收割
propagule 英 [ˈprɒpəgjuːl] 美 [ˈprɒpəgjuːl] n. 能发育成植物体的芽；繁殖体
variation 英 [ˌveərɪˈeɪʃn] 美 [ˌverɪˈeɪʃn] n. 变化；变异
clonal line 无性系植物
provenance 英 [ˈprɒvənəns] 美 [ˈprɑːvənəns] n. 起源；出处
provenance study 产地研究
nursery 英 [ˈnɜːsəri] 美 [ˈnɜːrsəri] n. 苗圃；温床；滋生地
target plant 目标植物
warrant 英 [ˈwɒrənt] 美 [ˈwɔːrənt] vt. 保证

Useful Expressions

1. derivative 英 [dɪˈrɪvətɪv] 美 [dɪˈrɪvətɪv] n. 衍生物；派生物
This isn't an entirely new car, but a new derivative of the Citroen XM.
这不是一款全新的车,而是雪铁龙 XM 系列的派生车型。

2. identify 英 [aɪˈdentɪfaɪ] 美 [aɪˈdentəfaɪ] vt. 确定；识别；认出
There are a number of distinguishing characteristics by which you can identify a Hollywood epic.
好莱坞的史诗大片有着一些与众不同的特点。

3. optimal 英 [ˈɒptɪməl] 美 [ˈɑːptɪməl] adj. 最佳的；最优的；最理想的
Cases are known where trioecy or, more generally, heteroecy, provides for an optimal mating system.
众所周知,单全异株,或者更通俗一些说,雌雄杂生为最适交配打下基础。

Questions

1. What is the function of Camptotheca acuminata?

2. What is the purpose of the medicinal plant research program at the LSU Agricultural Center?

3. How many seed sources were collected in natural distribution? What are the uses of these collected seeds?

1.2　Species of Medicinal Plants Resources in Other Countries

Passage 1

Introduction of Medicinal Plants Species with the Most Traditional Usage in Alamut Region

Introduction

Before the introduction of chemical medicines, man relied on the healing properties of medicinal plants. Some people value these plants due to the ancient belief which says plants are created to supply man with food, medical treatment, and other effects. It is thought that about 80% of the 5.2 billion people of the world live in the less developed countries and the World Health Organization estimates that about 80% of these people rely almost exclusively on traditional medicine for their primary healthcare needs. Medicinal plants are the "backbone" of traditional medicine, which means more than 3.3 billion people in the less developed countries utilize medicinal plants on a regular basis. There are nearly 2,000 ethnic groups in the world, and almost every group has its own traditional medical knowledge and experiences. Iran is home to several indigenous tribes with a rich heritage of knowledge on the uses of medicinal plants. Iran has varied climates and geographical regions that have caused a wide distribution of individual medicinal plant species such that each tribe has its own plants and customs. Alamut is one of the most important geographic regions in Iran because of its ancient history of cultivating traditional medicinal plants. Alamut region and the several villages it encompasses are secluded from other cities in Iran, which is the reason why the people living in this region have relied on indigenous medical knowledge and medicinal plants. In this study, we analyzed the medicinal plants with most therapeutic usage in the region.

Alamut mountainous region is situated in the central Alborz Mountains, between $36° 24'$ and $36° 46'$ northern latitudes and $50° 30'$ and $50° 51'$ eastern longitudes with an altitude ranging from 2,140 to 4,175 m. The region is located on the northeast of Ghazvin Province and is bounded to the north by the Mazandaran Province in Tonekabon and bounded on the east by Tehran Province in the Taleghan mountains. Annually, it rains 368.03 mm and the average temperature is 14℃. Topography is distinctly marked with several mountains, springs, rivulets, and rivers. This area is geographically located in the

Irano-Turanian region.

Study Area: Iran Map and Alamut in Ghazvin Province

The ethnic composition of the region is quite diverse and almost 90% of its population resides in rural areas. The language of the inhabitants is known as Deylamite. People of Alamut have a long history of exporting medicinal plants to other regions of Iran. Roadways have increased communication among the rural natives in Alamut and have also increased tourism to the region because of its several ancient castles. Because of good quality of medicinal plants in this region and more immethodical pick of them, some of species have become extinct. For this reason, an important aim of this study is to protect the preservation of the region's plants. Other aims include the following:

Documenting the traditional knowledge of medicinal plants from the natives;

Assessing the most commonly used local medicinal plants;

Promoting the potential benefits of medicinal plants.

Data Collection

We first prepared a map with a scale of 1:25,000 from the region to identify the number of villages, roads, and vegetations. We visited the region and spoke to herbal practitioners and village seniors. A questionnaire was used to obtain information on the types of ailments treated using traditional medicinal plant species. Sometimes informants were asked to come to the field and introduce us to the plants. When this was not possible, plants were collected around the villages of the informants and were shown to them to confirm the plant names. This investigation took over 2 years and information was collected 1—2 days per week. Voucher samples were also collected for each plant and were identified using floristic, taxonomic references. Flora Iranica and a dictionary of Iranian plant names were used for identification purposes. Plants were deposited at the Herbarium of Institute of Medicinal Plants.

Results

Although ancient sages through trial and error methods have developed herbal medicines, the reported uses of plant species do not certify their efficacy. Reports on ethnomedicinal uses of plant species require pharmacological screenings, chemical analyses, and tests for their bioactive activities. Pharmacological screening of plant extracts provides insight to both their therapeutic and toxic properties as well as helps in eliminating the medicinal plants or practices that may be harmful.

This study provides information on 16 medicinal plants belonging to 12 families that are most commonly used for traditional medicine in Alamut region. Botanical names of plants were sorted alphabetically, and for each species and the following information was hence represented: family, vernacular name, part used. Traditional use and preparation was compared with other references.

Among these medicinal plants, Apiaceae, Lamiaceae, and Boraginaceae were the most dominant families with 4, 2, 2 species belonging to 4, 2, 2 genera of medicinal plants, respectively.

Of the 16 medicinal plants, 8 species had similar effects in traditional and medicinal uses when comparing Alamut with other references. Achillea millefolium had antibacterial effects; Capparis spinosa is used for headache, renal complaints and stimulating tonic; Echium amoenum is used for common cold and had sedative effects; Ferula persica is used for gout; Juglans regia is used for diabetes; Smyrnium cordifolium is edible and used as tonic; Viola odorata is used for fever and migraine; Ziziphora clinopodioides is used for cold, infections and stomachache.

Some effects which are mentioned in traditional medicine of Alamut region were important with no scientific information about them. For example, Berberis integerrima and Hippophae rhamnoides had good effect on lowering of serum lipids and blood sugar and hypertension. Malva neglecta is used for mouth fungal infection in children and Stachys lavandulifolia is used for headache and renal calculus. Other researches can perform experiments to discover their components and effects.

All of the medicinal plants were collected from the wild or in the native people's gardens. Some medicinal plants can no longer be found in the region and are only cultivated in the native people's gardens. For example, Echium amoenum is an endemic plants in Iran with historically wide spread in the region, but because of frequent picking, the species is now just cultivated in the native people's gardens.

Different parts of medicinal plants were used by the inhabitants of Alamut region as medicine for treating ailments. The most common parts used were flowers (25%). The uses of aerial parts, leaves, fruits and roots were the same (15%). The uses of the stems (7%), seeds, and blooms (4%) were lower than the others. The 16 medicinal plant species were used in treating 27 different types of ailment.

http://pubmedcentralcanada.ca/pmcc/articles/PMC3813099/

Vocabulary

indigenous 英 [ɪnˈdɪdʒənəs] 美 [ɪnˈdɪdʒənəs] *adj.* 土生土长的；本地的
therapeutic 英 [ˌθerəˈpjuːtɪk] 美 [ˌθɛrəˈpjutɪk] *adj.* 治疗(学)的；疗法的
northern latitude 北纬
eastern longitude 东经
altitude 英 [ˈæltɪtjuːd] 美 [ˈæltɪtuːd] *n.* 高度；海拔高度；[天]地平纬度
rivulet 英 [ˈrɪvjələt] 美 [ˈrɪvjəlɪt] *n.* 小河；小溪
ethnic composition 民族构成；种族构成；民族成分
questionnaire 英 [ˌkwestʃəˈneə(r)] 美 [ˌkwestʃəˈner] *n.* 调查问卷；调查表
ancient sages 先贤
ethno 英 [ˈeθnəʊ] 美 [ˈeθnoʊ] *n.* 民族

pharmacological 英 [ˌfɑːməkəˈlɒdʒɪkl] 美 [ˌfɑːməkəˈlɒdʒɪkl] *adj.* 药理学的
bioactive 英 [ˌbaɪˈæktɪv] 美 [ˌbaɪˈæktɪv] *adj.* 对活质起作用的；生物活性的
extract 英 [ˈekstrækt] 美 [ɪkˈstrækt] *n.* 汁；摘录；提炼物；浓缩物
toxic 英 [ˈtɒksɪk] 美 [ˈtɑːksɪk] *adj.* 有毒的
vernacular 英 [vəˈnækjələ(r)] 美 [vərˈnækjələ(r)] *n.* 动植物的俗名
renal 英 [ˈriːnl] 美 [ˈrinəl] *adj.* 肾脏的
tonic 英 [ˈtɒnɪk] 美 [ˈtɑːnɪk] *n.* 滋补品；奎宁水
sedative 英 [ˈsedətɪv] 美 [ˈsɛdətɪv] *n.* [医] 镇静剂；止痛药
Echium amoenum 无芒雀麦
inhabitant 英 [ɪnˈhæbɪtənt] 美 [ɪnˈhæbɪtənt] *n.* 居民；（栖息在某地区的）动物

Useful Expressions

1. utilize 英 [ˈjuːtəlaɪz] 美 [ˈjutlˌaɪz] *vt.* 利用；使用
Sound engineers utilize a range of techniques to enhance the quality of the recordings.
音响师运用一系列技术来提高录音质量。

2. seclude 英 [sɪˈkluːd] 美 [sɪˈklud] *vt.* 使隔开；使隔绝；使隐退
It was not right to seclude themselves like that.
他们不应当那样与世隔绝。

3. alphabetically 英 [ˌælfəˈbetɪklɪ] 美 [ˌælfəˈbetɪklɪ] *adv.* 照字母顺序排列地
They are all filed alphabetically under author.
这些都是按照作者姓名的字母顺序归档的。

4. reside 英 [rɪˈzaɪd] 美 [rɪˈzaɪd] *vi.* 住；居住
Margaret resides with her invalid mother in a London suburb.
玛格丽特同她病弱的母亲住在伦敦郊区。

5. confirm 英 [kənˈfɜːm] 美 [kənˈfɜːrm] *vt.* 证实；[法] 确认
X-rays have confirmed that he has not broken any bones.
X 光片证实他没有骨折。

6. certify 英 [ˈsɜːtɪfaɪ] 美 [ˈsɜːrtɪfaɪ] *vt.* 证明；颁发（或授予）专业合格证书
The National Election Council is supposed to certify the results of the election.
全国选举委员会应对选举结果出具证明。

Questions

1. Why does the author choose Alamut as study area?
2. How are the data collected?
3. What does the study get?

Passage 2

Medicinal Plants List Ten Common Species in North America

The following is a medicinal plants list of common species that can be found in most areas throughout North America.

Nature has blessed us with an array of amazing medicinal plants. These plants can be found right outside our doorstep from the spunky, dominating dandelion to the handsome stalks of stinging nettle.

Plants can be used in many ways to improve overall health and wellness. Plants can give us the power to take control of our own health, so that we may be the best, most vibrant versions of ourselves!

The plants I have included in this medicinal plants list is a lovely place to start! I've included a few medicinal uses for each to help inspire you to get to know them further:

Dandelion (*Taraxacum officinale*)

Parts Used: Leaves, Flower, Roots

Medicinal Uses: All plant parts when taken internally can be a digestion ally, mild laxative, can support the body's ability to absorb nutrients, and provide liver support.

Safety Issues: For some, dandelion can be a powerful diuretic when taken in high doses or too frequently. Avoid if you have a latex allergy.

Plantain (*Plantago major*)

Parts Used: Leaves

Medicinal Uses: Externally, crushed leaves can treat many bug bites, bee stings, and even nettle rashes by reducing inflammation and pain. It also can be used internally, as a tea, for inflammation due to excessive coughing.

Stinging Nettle (*Urtica dioica*)

Parts Used: Leaves

Medicinal Uses: Externally, a strong tea can be used as a hair rinse for oily hair, and aids in the treatment of eczema. Drinking the tea is a great herbal ally when treating anemia and pollen allergies.

Safety Issues: Caution-stinging hairs located on stalk and underneath leaves have formic acid within them. When touched it can cause an uncomfortable rash, painful itch or small bumps on affected area. Use gloves when interacting with fresh nettle.

Burdock (*Arctium minus*)

Parts Used: Roots, Leaves

Medicinal Uses: It can help create an appetite and aid in overall digestion. Internally and externally it can be used for an array of skin problems such as dry skin, eczema or cracked skin.

Yarrow (*Achillea millefolium*)

Parts Used: Leaves, Roots, Flowers

Medicinal Uses: Used internally, the tea reduces fevers, can lower blood pressure, is anti-inflammatory, and aids with symptoms of diarrhea. Externally, it slows bleeding when applied topically to wounds in the form of a poultice.

Safety Issues: Should not be taken extensively when pregnant.

Self-Heal (*Prunella vulgaris*)

Parts Used: Flowers, Leaves

Medicinal Uses: Externally, it aids in the healing of cuts, sores and any open wounds. It also has anti-inflammatory properties.

Blackberry (*Rubus spp.*)

Parts Used: Leaves, Roots, Berries

Medicinal Uses: Tea made from bark and leaves, used internally, can aid in symptoms of diarrhea and inflammation. Chewing leaves, bark, or gargling tea can aid in mouth or throat irritations.

Willow (*Salix spp.*)

Parts Used: Bark, Leaves

Medicinal Uses: Anti-inflammatory, nature's aspirin, eases pain both internally and externally, reduces fever, can help boost the immune system, aids in menstrual pains, helps relieve headaches.

Safety Issues: This plant contains salicin—if you are sensitive to aspirin avoid using willow.

Echinacea (*Echinacea angustifolia*)

Parts Used: Flower, Leaves, Roots, and Seeds

Medicinal Uses: An immune booster, anti-microbial, will fight viral and bacterial infections, great for helping fight colds.

Chickweed (*Stellaria media*)

Parts Used: All above ground parts

Medicinal Uses: Externally used to aid in the healing of cuts and wounds; can sooth dry, itchy or irritated skin; and a poultice can be made to calm inflammation. Internally it aids in constipation, obesity, stomach, and bowel problems.

Please note that although many of the species in this common medicinal plants list are considered safe, it is always best to cross reference and be sure of both the identification and the properties of each plant you choose to harvest by consulting with multiple field guides and an experienced herbalist.

What Are You Waiting For!

Get up, get out and see if you can find some of these common plants! These are great plants to start your learning journey towards making wild medicines and eating wild foods! Working with plants, both cultivated and wild, is a way to start taking a mindful approach to our own health and wellness.

https://www.wildernesscollege.com/medicinal-plants-list.html

Vocabulary

array 英 [əˈreɪ] 美 [əˈreɪ] *n.* 数组；队列；阵列
spunky 英 [ˈspʌŋki] 美 [ˈspʌŋki] *adj.* 充满勇气的；精神十足的；灿烂的
dandelion 英 [ˈdændɪlaɪən] 美 [ˈdændlˌaɪən] *n.* [植] 蒲公英
nettle 英 [ˈnetl] 美 [ˈnetl:] *n.* 荨麻
diuretic 英 [ˌdaɪjuˈretɪk] 美 [ˌdaɪəˈretɪk] *n.* [医] 利尿剂
latex allerg 胶乳过敏；过敏症
plantain 英 [ˈplæntɪn] 美 [ˈplæntən] *n.* 车前草
stinging 英 [ˈstɪŋɪŋ] 美 [ˈstɪŋɪŋ] *adj.* 刺一样的；刺人的
ally 英 [ˈælaɪ] 美 [ˈælaɪ] *n.* 联盟；助手；支持者
anemia 英 [əˈniːmiə] 美 [əˈnimiə] *n.* 贫血症；无活力
pollen allergy 花粉变态反应
burdock 英 [ˈbɜːdɒk] 美 [ˈbɜːrdɑːk] *n.* 牛蒡属
eczema 英 [ˈeksɪmə] 美 [ɪgˈziːmə] *n.* [医] 湿疹
yarrow 英 [ˈjærəʊ] 美 [ˈjæroʊ] *n.* 西洋蓍草
diarrhea 英 [ˌdaɪəˈrɪə] 美 [ˌdaɪəˈriə] *n.* 腹泻
poultice 英 [ˈpəʊltɪs] 美 [ˈpoʊltɪs] *n.* 膏状药
sore 英 [sɔː(r)] 美 [sɔr, sor] *n.* （肌肤的）痛处；伤处
inflammation 英 [ˌɪnfləˈmeɪʃn] 美 [ˌɪnfləˈmeʃən] *n.* [医] 炎症
irritation 英 [ˌɪrɪˈteɪʃn] 美 [ˌɪrɪˈteʃən] *n.* 刺激
boost the immune system 增强免疫系统
menstrual 英 [ˈmenstruəl] 美 [ˈmɛnstruəl] *adj.* 月经的；每月的
salicin 英 [ˈsælɪsɪn] 美 [ˈsæləsɪn] *n.* 水杨苷
Echinacea 英 [ˌekəˈneɪʃiə] 美 [ˌɛkəˈnesiə, -ˈneʃə] *n.* 海胆亚目
immune booster 免疫增强剂
amicrobial 英 [maɪˈkrəʊbɪəl] 美 [maɪˈkroʊbɪrl] *adj.* 微生物的；由细菌引起的
viral 英 [ˈvaɪrəl] 美 [ˈvaɪrəl] *adj.* 病毒的；病毒引起的
constipation 英 [ˌkɒnstɪˈpeɪʃn] 美 [ˌkɑːnstɪˈpeɪʃn] *n.* [医] 便秘
bowel 英 [ˈbaʊəl] 美 [ˈbaʊəl, baʊl] *n.* 肠；内部
herbalist 英 [ˈhɜːbəlɪst] 美 [ˈɜːrbəlɪst] *n.* 种草药的人；草药商；草药医生
wellness 英 [ˈwelnəs] 美 [ˈwelnəs] *n.* 健康

Useful Expressions

1. bless 英 [bles] 美 [blɛs] *vt.* 祝福；保佑
"Bless you, Eva," he whispered.
"愿上帝保佑你，伊娃。"他低声说。

2. dominate 英 [ˈdɒmɪneɪt] 美 [ˈdɑːmɪneɪt] *v.* 支配；影响；在……中具有最重要的位置
The book is expected to dominate the best-seller lists.
这本书预计会占据畅销书排行榜的榜首。

3. vibrant 英 [ˈvaɪbrənt] 美 [ˈvaɪbrənt] *adj.* 振动的；响亮的；充满生气的
Tom felt himself being drawn towards her vibrant personality.
汤姆感觉自己被她充满朝气的个性所吸引。

4. internally 英 [ɪnˈtɜːnəlɪ] 美 [ɪnˈtɜːnəlɪ] *adv.* 内地；内心地；国内地；本质地
A great deal has been done internally to remedy the situation.
已经做了很多内部工作对这种情形进行补救。

5. soothe 英 [suːð] 美 [suð] *vt.* 安慰；缓和；使平静；减轻痛苦
He would take her in his arms and soothe her.
他会把她揽在怀中抚慰她。

6. externally 英 [ɪkˈstɜːnəlɪ] 美 [ɪkˈstɜːnlɪ] *adv.* 在（或从）外部
They can also be observed and used externally.
它们都能在外部观测和使用。

7. cultivate 英 [ˈkʌltɪveɪt] 美 [ˈkʌltəˌvet] *vt.* 耕作；种植；栽培
She also cultivated a small garden of her own.
她还耕种了一片属于自己的小园子。

Questions

1. What functions do plants possess?
2. Where can we find the useful plants?
3. Are medicinal plants absolutely safe?
4. What parts of medicinal plants are used as medicine?
5. What plants are used to deal with skin problems?

Chapter Two

Distribution of Medicinal Plants

2.1 Distribution of Medicinal Plants Resources in the World

Passage 1

Predicting the Distributions of Egypt's Medicinal Plants and Their Potential Shifts Under Future Climate Change (Part 1)

Introduction

Climate is one of the main elements used to describe plant niches, and climate change is considered one of the major threats to biodiversity. One of the main challenges today is to forecast how the various climate-change scenarios might affect species and communities in the future. To reduce the effects of climate change on biodiversity, we should be taking precautionary measures, in the long term to reduce emission gases, and in the short term to design appropriate networks of protected areas (PAs) for conservation.

Here we apply species distribution models (SDMs) to predict the impact of climate change on shifting distributions of Egyptian medicinal plants, and to evaluate the effectiveness of the Egyptian network of reserves. Medicinal plants are the main source of new herbal products and medicines, and therefore they are important for their roles in human health, the economy especially of poor areas, and culture and heritage. Between 70% and 80% of people across the world use medicinal plants as their traditional and primary system of health care. Because of the growing demand for natural health products and herbal medicines, the use of medicinal plants is increasing; in 1999 their annual market value was evaluated at $US 20 billion—40 billion, with an annual growth rate of 10%—20%. About 50,000—80,000 plant species are used for medicinal purposes across the world, with very variable distributed and mostly harvested from the wild. Their current extinction rate is 100 to 1,000 times higher than the natural background extinction rate, implying the loss of at least one important drug every two years. It is therefore very important to conserve these kinds of plants, by investigating their ecological requirements and increasing the awareness of decision makers, stakeholders and the public.

SDMs create a model of the relationship between species distribution and environmental predictors (such as climate, land cover/use, and habitat), which can then be projected under different climate scenarios. Modelling techniques rather than field data are more effective in evaluating the spatial reaction to climate drivers over large spatial and temporal scales. SDM methodology has developed very quickly so that currently there are many algorithms, platforms and software available. In our case study of the data-sparse country of Egypt, we have tried to keep the modelling as simple as possible, and hence we use just one SDM method (MaxEnt) with proven abilities in this field, especially in its ability to produce reasonable models with sparse data. The majority of countries find themselves in the position of having to make decisions about conservation in the face of climate change, but with few data to guide them.

SDMs are also used to assess the likely future value of PAs, important fundamental units in conservation which cover 14.8% of the entire land area of the world. There is a big effort to enlarge the coverage of PAs to 17% by 2020. The important question is to what extent these PAs can preserve species under climate change, recognising that many species will shift their distributions. Current PAs might still achieve significant roles if they provide suitable areas for species to disperse into new districts. The best strategy to preserve biodiversity is to recognise and protect the best locations—but are PAs in the "best" locations? There are several studies that measure current species richness inside and outside PAs; some found that species richness is higher inside, while others found the opposite. In this case study, we use species richness as our criterion to show the effectiveness of Egypt's network of PAs, given that the preservation of "biodiversity" is their objective in nearly all cases.

We focussed on medicinal plants because (i) validated dataset exists, unlike for all plants; and (ii) they are subjected to unregulated harvesting within Egypt by pharmaceutical companies, and hence there is more concern about them than other plants. Thus we evaluated the pattern of species richness for 114 Egyptian medicinal plants projected into the future to ask what the predicted patterns of diversity, gains, losses and turnover are under the various climate-change and dispersal scenarios. Then we ask whether Egypt's PAs provide and will provide suitable plant habitat compared with outside the PAs. There are about 30 PAs in Egypt covering approximately 15% of Egypt's land area. PAs are imperfect in their coverage of biodiversity, but because of their extent Egyptian PAs at least in theory protect biodiversity much better than those of many other countries. In all this work, "species richness" is understood to mean a quantitative estimate of the collective habitat suitability of all the 114 species combined, since that is what the raw MaxEnt output actually means.

Materials and Methods

Egypt has a total of 2,174 recorded species. Botanists have judged 121 of them as known to be "medicinal", but these are not representative of all Egypt's plants, with certain families (e. g. Chenopodiaceae, Labiatae, Zygophyllaceae) greatly over-represented, and some (e. g. Graminae, Leguminosae-Papilionoideae) greatly under-represented. The data were collated by the BioMAP project in Cairo between 2004—2008, funded by Italian Debt Swap; the data are presence-only, collected from different sources (i. e. literature, herbarium, and field work), but mainly recent field work especially in Sinai. There are virtually no available validated records of these species from surrounding countries, so we were unable to model the entire range of non-endemic species; this will often be the case for developing countries. All the data were examined in detail by a panel of expert field botanists during the BioMAP project, filtering out all records that were dubious on a variety of grounds (mainly identification and distribution). Much of the data derive from recent (since 1990) expeditions of academic botanists using GPS recorders, particularly to the wadis of Sinai, where plots were surveyed at approximately regular distances apart. Earlier records were carefully checked for the accuracy of the locations, and records with vague data removed. We removed species with fewer than ten spatially separate records, and those where the records were spatially very close together and with mean AUC of the models less than 0.5 (all to avoid overfitting). We eventually used 114 species consisting of 14,396 point records.

The environmental variables were 23 predictors, 19 of them (bioclimatic variables) downloaded from the WorldClim v1.4 dataset at a resolution of 2.5 arc-minutes (http://www.worldclim.org/bioclim). The Normalized Difference Vegetation Index (NDVI) data for seven years (2004 to 2010) were downloaded from the SPOT Vegetation website (http://free.vgt.vito.be/). These data are made available as maps synthesized over 10-day periods at resolution of 1 km. 252 maps represent data from 2004—2010, which were then clipped to Egypt's boundaries and used to create two predictors—maximum (Max_NDVI, indicating the maximum amount of vegetation there is per pixel) and the difference between the Minimum and Maximum (NDVI_differences, indicating the variability in vegetation per pixel). The maps were then rescaled to 2.5 arc minutes. The predictor layer of "habitat" was created by BioMAP, dividing Egypt into eleven classes (sea, littoral coast, cultivation, sand dune, wadi, metamorphic rock, igneous rock, gravels, serir sand sheets, sabkhas and sedimentary rocks). Altitude data were downloaded and rescaled to a pixel size of 2.5 arc-minutes.

Eleven of these 23 predictors were eventually used after removing collinearity by applying the Variance Inflation Factor using R v2.15. The predictors with the maximum variance inflation factor (VIF) values were discarded first, and new VIF scores calculated among the remaining predictors; this was repeated until all VIF scores were under 10.

Models for the current distribution of each species were projected into the future at the standard three different time slices (2020, 2050, and 2080) using predicted future climates from the Intergoverment Panel on Climate Change (IPCC)'s 4th assessment taken from the International Centre for Tropical Agriculture website (see http://www.ccafs-climate.org/), selecting two HadCM3 scenarios (A2a and B2a). Such global circulation models are widely used in SDMs to explore the effect of climate change on biodiversity, including likely shifts in distribution, habitat change, gains and losses, turnover and extinction. We chose to use the IPCC's 4th assessment, rather than the latest 5th assessment and its very different scenarios, for continuity with previous work and because the differences in SDMs are slight.

The A2 and B2 scenarios are regularly used in climate-change assessments. Both scenarios have different assumptions about the levels of CO_2 emissions. The A2 scenario expects this level to increase without barriers because of high growth rate in human population, not much technological development, expanded land-use change, and less environmental awareness. The B2 scenario expects the level not to change much because human population growth will be slower, with fewer changes in land-use, people are more environmentally conscious, and there is increasing technological invention. In this study, we do not take into account any phenological and/or evolutionary reactions to climate change, and hence we assume that species will attempt to find their climatically suitable habitat dependent on their dispersal capability. We are forced to make the assumption that some predictors (e.g. habitat, NDVI) do not change because there is no information or scenarios on how they might change; others (e.g. altitude) will clearly not change.

Maximum Entropy (MaxEnt) version 3.3.3k was used to run the models, choosing the options that created the best models (i.e. feature classes QPT, 10,000 background points, 1,000 iterations, cross-validation [$k=10$] with 10 replicates (for estimating prediction errors), 10% training presence threshold, and logistic output format, and to create the binarized (via thresholding) maps. MaxEnt performance is good with presence-only data and relatively few records. The option set was chosen after extensive tests to maximise the AUC and TSS, the standard measures of goodness-of-fit for SDMs. The one species with a mean AUC score less than 0.7 was removed from the analysis, as recommended. Following, we chose the "10% training presence" threshold rule to give a binary map for each of the 10 replicates for each species. With the 10% training presence rule, the threshold selects pixels where 90% of the training presences are correctly classified, thus giving some allowance for the possibilities of ephemeral populations or recording errors. A minimal training presence threshold that correctly predicts every training presence may lead to over-prediction. The resulting maps were analysed under the two extreme dispersal assumptions (unlimited and no-dispersal) usually applied in such studies. Both assumptions have been criticized for not involving the impact of biotic interactions on species distributions, but there is no practical way of incorporating any putative interactions for Egyptian species. Species gains occur when species occupy new areas, while losses can occur when the habitat suitability is reduced for that area in the future. Species turnover is the differences between the current and future species composition of the same location, used as an indicator at large or regional scales.

MaxEnt does not automatically produce a consensus binary (presence/absence) map from the 10 replicate runs, and hence this was made manually using the Raster Calculator and Reclassify tools in ArcGIS 10.2.2. For unlimited dispersal, the consensus map allotted a "presence" to a pixel that had presence values in more than 50% of the model runs (i.e. >5 replicates). Under the no-dispersal assumption, the consensus maps for the two time periods (e.g. "current" and "2020") were compared so as to allocate a pixel to be a "presence" in the event that it was a "presence" in both maps.

Maps for the distribution of species richness for current and future times were produced by summing all predicted consensus distribution maps for the individual species together (assuming either unlimited or no-dispersal in the future). Thus, maps were produced for species richness for future times (2020, 2050, and 2080) under both scenarios (A2 and B2) for both unlimited and no-dispersal assumptions, a total of 12 species-richness maps. Change maps were produced to aid interpretation by subtracting one map from another.

We then compared the species richness inside and outside PAs to see to what extent they protect Egyptian plants now, and in the future, using the oldest and largest 25 PAs, excluding just the most recent and smallest. We chose 2,000 pixels at random, and then created a 50-km buffer around each PA. We chose 50-km as the tradeoff between a small value (to try to ensure the habitat outside is as similar as possible) and a figure large enough to ensure enough of the random pixels lay within it. The mean species richness was then calculated for the random pixels lying within each PA, and lying outside but within the appropriate buffer zone, creating paired values inside and outside of each PA. In the case of very small PAs which are smaller than one pixel (whose side is 4.64 km), species richness was taken to be the value for the pixel containing the PA. The paired difference inside-outside was calculated for each PA, and the mean difference tested in a one-sample t-test against the null hypothesis that the mean equals zero.

To estimate how much each species loses or gains in habitat suitability, the number of pixels gained and lost for each species was calculated, and the gain maps (future times × scenarios emissions) were summed together across species to produce a "gain in suitability" map to demonstrate which locations are projected to gain more species in the future; sites can only gain species under the assumption of unlimited dispersal, and such sites may constitute potential new PAs in the future. A loss map shows locations predicted to lose suitability for species in the future (assuming either unlimited and no-dispersal).

Species turnover is defined as the number of species changes in specific locations, used as a suitable measure of change in species composition to evaluate the effect of climate change on biodiversity from regional to continental levels. It is calculated as the differences between present-day and future species composition. Locations with small turnover are predicted to remain suitable in the future, whereas locations with higher turnover indicate less stable habitat suitability with time. Species turnover for unlimited dispersal was calculated as SG + SL, where SG is the number of species gained, and SL is the number of species lost. The equation normally expresses turnover as a proportion of the species richness of each pixel, but this was unsuitable in the case of Egyptian plants because of the very low richness in areas of the western desert, which made some of the proportional turnover values unreasonably large. Species turnover for no-dispersal is just equal to species losses because there can be no gains.

http://journals.plos.org/plosone/article?id=10.1371/journal.pone.0187714

Vocabulary

niches 英 [nɪtʃ] 美 [niːʃ] *n.* 商机
precautionary 英 [prɪˈkɔːʃənrɪ] 美 [prɪˈkɔʃəˌneri] *adj.* 预先警戒的；小心的
emission 英 [iˈmɪʃn] 美 [ɪˈmɪʃən] *n.* 排放；辐射；排放物；散发物（尤指气体）
extinction 英 [ɪkˈstɪŋkʃn] 美 [ɪkˈstɪŋkʃən] *n.* 熄灭；消灭；灭绝
temporal 英 [ˈtempərəl] 美 [ˈtɛmpərəl, ˈtɛmprəl] *adj.* 时间的；世俗的；暂存的
methodology 英 [ˌmeθəˈdɒlədʒɪ] 美 [ˌmɛθəˈdɑːlədʒi] *n.* 方法论；一套方法
algorithm 英 [ˈælɡərɪðəm] 美 [ˈælɡəˌrɪðəm] *n.* 演算法；运算法则；计算程序
dune 英 [djuːn] 美 [duːn] *n.* （由风吹积成的）沙丘
wadi 英 [ˈwɒdi] 美 [ˈwɑːdi] *n.* 旱谷；干涸河道；溪流；多岩石的干涸河床
metamorphic 英 [ˌmetəˈmɔːfɪk] 美 [ˌmetəˈmɔːrfɪk] *adj.* 变质的；改变结构的
igneous 英 [ˈɪɡnɪəs] 美 [ˈɪɡnɪəs] *adj.* （尤指岩石）火成的
scenario 英 [səˈnɑːriəʊ] 美 [səˈnæriˌoʊ] *n.* （行动的）方案；剧情概要
dispersal 英 [dɪˈspɜːsl] 美 [dɪˈspɜːrsl] *n.* 分散；散布；消散；驱散
consensus binary 共识二进制
pixel 英 [ˈpɪksl] 美 [ˈpɪksəl, -ˌsɛl] *n.* （显示器或电视机图像的）像素
turnover 英 [ˈtɜːnəʊvə(r)] 美 [ˈtɜːrnoʊvə(r)] *n.* 翻滚；逆转；转向；成交量

Useful Expressions

1. imply 英 [ɪmˈplaɪ] 美 [ɪmˈplaɪ] *vt.* 暗示；意味；隐含
"Are you implying that I have something to do with those attacks?" She asked coldly.
"你在暗示我和那些袭击有关吗？"她冷冷地问。

2. enlarge 英 [ɪnˈlɑːdʒ] 美 [ɪnˈlɑːrdʒ] *vt.* 扩大；扩展
The glands in the neck may enlarge.
颈部腺体可能增大。

3. maximize 英 [ˈmæksɪmaɪz] 美 [ˈmæksəˌmaɪz] *vt.* 最大化；最大限度利用（某事物）
In order to maximize profit, the firm would seek to maximize output.

为了获得最大利润,这家公司会把产量增至最大。

Questions

1. What should we do to reduce the effects of climate change on biodiversity?
2. Why SDMs are used in this study?
3. How many species are medicinal in Egypt according to botanists?
4. What is species turnover?

Passage 2

Predicting the Distributions of Egypt's Medicinal Plants and Their Potential Shifts Under Future Climate Change (Part 2)

Results

The performance of the models was good in term of AUC (0.90 ± 0.004, 0.80 to 0.98) and TSS scores (0.63 ± 0.01, 0.27 to 0.85); model performance was not affected by sample size. Three of the temperature-related variables had the most important effect on predicting distributions.

The pattern of species richness increased toward the north and east of Egypt, especially the Mediterranean coast and Sinai. With unlimited dispersal, binary species richness was predicted to increase for all future times (2020, 2050, and 2080) and both scenarios (A2 and B2). Using binary distributions, the pattern of species richness was predicted under both dispersal assumptions. For unlimited dispersal, by 2020 under the A2 scenario species richness is predicted to increase along the Mediterranean coast and in the Qattara Depression, and in scattered areas on both sides of the Suez Gulf, the Red Sea coast, and Sinai. These predicted increases become more obvious by 2050 and very marked by 2080. Decreases in species richness are predicted to occur in small scattered areas, similar in all future time periods. There were clear differences under the B2 scenario; by 2020 the predicted increases are lower, and include some strong predicted declines along both sides of the Suez Gulf and further south along the Red Sea coast.

By 2050 species richness is predicted to increase generally, eliminating the previous declines. By 2080 the increases are predicted to be more marked in the same areas. Assuming no-dispersal, the predicted pattern under the A2 scenario shows no change or slight declines by 2020, more marked on both sides of Suez Gulf, along the Red Sea coast, and some scattered areas around greater Cairo. By 2050, these predicted declines are less except for areas around greater Cairo. By 2080, predicted species richness declines are only marked on both sides of the Suez Gulf, the Red Sea coast, and in central Sinai and greater Cairo. Under the B2 scenario, the predicted pattern of decline is similar but the magnitudes are greater except for 2080.

The overall average predicted mean species richnesses for binary distributions, assuming unlimited dispersal, gradually increase for future times (2020, 2050, and 2080) under both A2 and B2 scenarios compared to the present day, while assuming no-dispersal results in predicted declines for both scenarios compared with the present day.

The predicted species richness is significantly higher inside PAs than outside in all combinations of assumptions (overall mean excess is 10.2 ± 0.6; a one-sample t-test shows this is significantly greater than zero—$t323=10.1$, $p<0.001$). The overall mean species richness outside the PAs is 39.1, and hence on average there is a 26% increase inside the PA.

For unlimited dispersal, the mean predicted species-richness excess inside relative to outside PAs declined slightly with time. With no dispersal, this predicted excess inside relative to outside declined slightly under both scenarios.

With unlimited dispersal, under the A2 scenario the maximum species gain by 2020 was predicted to be at the Mediterranean coast, along both sides of the Suez Gulf, along the Red Sea coast further to the south, and some scattered patches in Sinai. By 2050, the highest predicted gains appeared along the Mediterranean coast and the Qattara Depression, scattered places on both sides of the Suez Gulf and the Red Sea coast. By 2080, the highest species gains were predicted to be along the Mediterranean coast, the Qattara Depression to Wadi El-Natrun, small areas around greater Cairo, and scattered areas along the Red Sea coast. There were no big differences in predicted gains under the B2 scenario, although the magnitudes were lower.

For unlimited dispersal, under the A2 scenario the maximum species losses by 2020 were predicted to be on both sides of the Suez Gulf, along the Red Sea coast, and some small scattered areas in Sinai, east of Cairo and around Ismailia. By 2050, they were predicted to be along the Red Sea coast and areas around greater Cairo, and by 2080, along the Mediterranean coast, north and south Sinai, some scattered areas on both sides of the Suez Gulf, and along the Red Sea coast. Under the B2 scenario, predicted losses were higher but similar in pattern.

For the no-dispersal assumption, under both scenarios the pattern of predicted losses did not differ much from those for unlimited dispersal. Differences were only apparent in the magnitudes, which were lower under the no-dispersal assumption.

For unlimited dispersal, under the A2 scenario by 2020 the highest turnover was predicted to occur along the Red Sea coast, in Sinai, along the Mediterranean coast, and along both sides of the Suez Gulf. By 2050 the pattern is similar but the magnitude is lower. By 2080, the highest predicted species turnover was along the Mediterranean coast, greater Cairo southwards, scattered areas both sides of the Suez Gulf, and some scattered areas along the Red Sea coast. Under the B2 scenario, the pattern of turnover was very similar, with some differences in magnitude. Under the no-dispersal assumption, turnover for both scenarios and all times were also similar to the pattern of losses under the no-dispersal assumption.

http://journals.plos.org/plosone/article?id=10.1371/journal.pone.0187714

Vocabulary

Mediterranean coast   地中海海岸；地中海沿岸
eliminate 英 [ɪˈlɪmɪneɪt] 美 [ɪˈlɪməˌnet] vt. 淘汰；排除；消除
magnitude 英 [ˈmæɡnɪtjuːd] 美 [ˈmæɡnɪtuːd] n. 量级；巨大；重大

Useful Expressions

1. scatter 英[ˈskætə(r)] 美 [ˈskætɚ] vt. （使）散开；驱散
She tore the rose apart and scattered the petals over the grave.
她掰开玫瑰花，将花瓣撒在坟墓上。
2. assume 英 [əˈsjuːm] 美 [əˈsuːm] v. 承担；呈现；假定；认为
It is a misconception to assume that the two continents are similar.
关于这两块大陆相似的假设是一种误解。

Questions

1. Please give a brief account of the result.
2. Are there any differences in predicted gains under the B2 scenario?
3. Where is the highest predicted species turnover?

2.2　Natural Distribution of Medicinal Plants Resources in China

Passage 1

Colonization and Diversity of AM Fungi by Morphological Analysis on Medicinal Plants in Southeast China (Extract)

Introduction

Arbuscular mycorrhizal (AM) fungi, the most ubiquitous symbiosis in nature, are a kind of these soil microbes. Reports suggest that estimated 80% of plant species forms mycorrhizas. In general, AM fungi and the host plants are reciprocal symbionts. The symbiosis improves plants the nutrient uptake and provides protection from pathogens, while the AM fungi receive carbohydrates.

All over the world, 80% of the rural population in developing countries utilizes locally medicinal plants for primary healthcare. And in China, the use of different parts of medicinal plants to cure specific illness has been popular from ancient time. In Zhangzhou, southeast China, the typical humid subtropical monsoon climate contributes to the growth of more than 700 kinds of lush medicinal plants and creates unique ecological conditions for species diversity and distribution of AM fungi.

The distribution of AM fungi associated with medicinal plants has been reported. In a survey on AM association with three different endangered species of Leptadenia reticulata, Mitragyna parvifolia, and Withania coagulans, high diversity of AM fungi was observed, and Glomus constrictum, Glomus fasciculatum, Glomus geosporum, Glomus intraradices, Glomus mosseae, and Glomus rubiforme were the most dominant species. Similarly, 34 AM fungal species were identified from 36 medicinal plant species. Approximately 15 fungal species from 10 genera were isolated from the collected soils in medicinal plant species, lemon balm (Melissa officinalis L.), sage (Salvia officinalis L.), and lavender (Lavandula angustifolia Mill). About 50 species of medicinal plants from 19 families have been studied in the association with AM fungi.

However, not enough has been focused on the mycorrhizal association with medicinal plants. Generally, AM fungi species in different ecosystems are affected by edaphic factors, so it is necessary to investigate the spatial distribution and colonization of AM fungi related to the medicinal plants. Hence, the present study is attempted to investigate the diversity of AM fungi associated with medicinal plant species in Zhangzhou, southeast China.

Materials and Methods

1) Study Sites

The city of Zhangzhou, Fujian Province, a subtropical region, is located on 23°08′—25°06′N and 116°53′—118°09′E. The mean annual temperature is 21℃ with yearly precipitation of 1,000—1,700 mm and annual sunshine of 2,000—2,300 hours. Frost-free periods add up to more than 330 days with cool summer and warm winter. The medicinal plants in this study were collected from Xiaoxi Town (24°44′N, 118°17′E), which was cinnamon soil from farmland, and Guoqiang village (24°35′N, 117°56′E), which was cinnamon soil from woodland, in Zhangzhou.

2) Sample Collection

The plants grew under natural environmental conditions. Six healthy individuals per plant species of medicinal plants were randomly selected for the collection of rhizospheric soil and root samples; 180 soil and root samples were collected from Xiaoxi Town and Guoqiang Village in October 2011. For each plant, three random soil cores at the depth of 0—30 cm about 1,000 g were established by contacting from the 6 duplicate plants. Approximately 20 plants species and 120 soil samples were collected in total. The subsamples were air-dried for 2 weeks and stored in sealed plastic bags at 4℃ for the following analysis.

3) Estimation of AM Colonization

The mixed soil and roots samples of each plant species were packed in polyethylene bags, labeled and brought to the laboratory. The soil samples were air-dried at room temperature. Roots were washed to remove soil particles, preserved with FAA. For colonization measurement, roots were cleared in 10% (w/v) KOH and placed in a water bath (90℃) for 20—30 minutes. The cooled root samples were then washed with water and stained with 0.5% (w/v) acid fuchsin. Fifty root fragments for each sample (ca. 1 cm long) were mounted on slides in a polyvinyl alcohol solution and examined for the presence of AM structures at 100—400x magnification with an Olympus BX50 microscope for the presence of AM structures. The percentage of root colonization was calculated.

4) AM Fungus Spore Quantification and Identification

Three aliquots of soil (20 g) were obtained for every plant species. AM fungal spores were extracted from the soil samples by wet sieving and sucrose density gradient centrifugation. Spores were counted under a dissecting microscope, and spore densities (SD) were expressed as the number of spores per 100 g of soil. The isolated spores were mounted in polyvinyl lactoglycerol (PVLG). Morphological identification of spores up to species level was based on spore size, color, thickness of the wall layers, and the subtending hyphae by the identification manual and the website of the international collection of vesicular and AM fungi (http://invam.wvu.edu/).

5) Soil Analysis

Soil samples were air-dried and sieved through 2 mm grid. Three rhizospheric soil samples (2 mm fraction) for each medicinal plant were analyzed for their pH, electrical conductivity (EC), organic matter (OM) content, available N (N), available P (P), and available K (K). Soil pH was measured in soil water suspension 1:2 (w/v) by pH meter. EC was measured at room temperature in soil suspension 1:5 (w/v) using conductivity meter. OM content was determined by the Walkley-Black acid digestion method. P (extracted with 0.03 M NH4F-0.02 M HCl) was measured by molybdenum blue colorimetry, K by an ammonium acetate method using a flame photometer, and N by the alkaline hydrolysis diffusion method.

6) Diversity Studies

Ecological measures of diversity, including spore density (SD), species richness (SR), isolation frequency (IF), Shannon-Wiener index, and evenness, were used to describe the structure of AM fungi communities. Diversity studies were carried out from Zhangzhou separately for abundance and diversity of AM fungal species. Spore density was defined as the number of AM fungi spores and sporocarps in 100 g soil; species richness was measured as the number of AM fungi species present in soil sample; isolation frequency (IF) = (number of samples in which the species or genus was observed/total samples) × 100%. Species diversity was assessed by the Shannon-Weiner index as follows; species evenness is calculated by the following formula: where = total number of species in the community (richness). It is the relative abundance of each identified species per sampling site and is calculated.

7) Statistical Analysis

The analysis of Pearson correlation coefficient, variance (ANOVA), and principal component were all carried out with SPSS Bass 18.0 (SPSS Inc., USA). The Pearson correlation coefficient was employed to determine the relationships between AM colonization, SD, SR, IF, and soil parameters. Differences in soil parameters, colonization, SD, SR, and IF were tested using one-way ANOVA and means were compared by least significant difference at 5% level.

https://www.hindawi.com/journals/tswj/2015/753842/

Vocabulary

arbuscular mycorrhizal　丛枝菌根；菌根真菌
ubiquitous 英 [juːˈbɪkwɪtəs] 美 [juˈbɪkwɪtəs] *adj.* 无所不在的；普遍存在的
symbiosis 英 [ˌsɪmbaɪˈəʊsɪs] 美 [ˌsɪmbaɪˈoʊsɪs] *n.* (通常互利的)共生
soil microbes　土壤微生物
mycorrhiza 英 [ˌmaɪkəˈraɪzə] 美 [ˌmaɪkəˈraɪzə] *n.* 菌根
reciprocal 英 [rɪˈsɪprəkl] 美 [rɪˈsɪprəkəl] *adj.* 互惠的；相互的
symbiont 英 [ˈsɪmbaɪɒnt] 美 [ˈsɪmbaɪˌɒnt] *n.* 共生有机体；共生体
nutrient 英 [ˈnjuːtriənt] 美 [ˈnuːtriənt] *n.* 营养物；养分；养料
uptake 英 [ˈʌpteɪk] 美 [ˈʌpˌtek] *n.* [术]摄入
pathogen 英 [ˈpæθədʒən] 美 [ˈpæθədʒən] *n.* 病菌；病原体
carbohydrate 英 [ˌkɑːbəʊˈhaɪdreɪt] 美 [ˌkɑːrboʊˈhaɪdreɪt] *n.* 碳水化合物；糖类；淀粉质或糖类食物
monsoon 英 [ˌmɒnˈsuːn] 美 [ˌmɑːnˈsuːn] *n.* 季风；季风雨；夏季季风
mycorrhizal fungi　菌根真菌
ecosystem 英 [ˈiːkəʊsɪstəm] 美 [ˈiːkoʊsɪstəm] *n.* [生]生态系统
edaphic 英 [iˈdæfɪk] 美 [iˈdæfɪk] *adj.* 土壤的
cinnamon 英 [ˈsɪnəmən] 美 [ˈsɪnəmən] *n.* 樟属植物；桂皮香料
duplicate 英 [ˈdjuːplɪkeɪt] 美 [ˈduːplɪkeɪt] *n.* 完全一样的东西；复制品
aliquots 英 [ˈælɪkwɒt] 美 [ˈælɪkwɑːt] *n.* 试样
fungal spore 英 [ˈfʌŋɡəl spɔː] 美 [ˈfʌŋɡəl spɔr] *n.* 真菌孢子
rhizospheric 英 [raɪˈzɒsferɪk] 美 [raɪˈzɒsferɪk] *adj.* [医] 根际的

Useful Expressions

1. colonization 英 [ˌkɒlənaɪˈzeɪʃn] 美 [ˌkɑːlənɪˈzeʃən] *n.* 殖民地的开拓；殖民；移殖

But it also supported the colonization by European.

但后来也一度支持了欧洲的殖民扩张。

2. sieve 英 [sɪv] 美 [sɪv] v. 筛；筛选；滤

Cream the margarine in a small bowl, then sieve the icing sugar into it.

将人造黄油在小碗里搅成糊状，再将糖粉筛进去。

3. dissect 英 [dɪˈsekt] 美 [ˈdaɪˌsekt] vt. 解剖；仔细分析

We dissected a frog in biology class.

我们在生物课上解剖了一只青蛙。

Questions

1. How many people in developing countries use locally medicinal plants for primary healthcare?

2. Why is it necessary to study the relationship between the colonization of AM fungi and the medicinal plants?

3. How is sample collected?

Passage 2

The Ability of Nature Reserves to Conserve Medicinal Plant Resources: A Case Study in Northeast China (Extract)

Discussion

Four years of intensive field surveys resulted in our collection of detailed information about the distributions of 49 medicinal plant species. We used Maxent to model the potential habitat distributions of the 49 medicinal plant species and found that northeast China is rich in medicinal plant resources and these plants are potentially influenced by negative environmental factors, such as climate change and human disturbance. In particular, we collected data on six endangered plants under national protection. The data we collected allowed us to then predict the potential distribution for each of these species. By identifying high-AUC values for each species, we were able to reliably predict the potential geographical distributions of these species, as opposed to using anecdotal reports, which may be less accurate for predicting the current distribution.

We found that the most important variables for predicting the distribution of these 49 species are annual mean precipitation and precipitation seasonality. As major drivers of habitat type, annual mean precipitation and precipitation seasonality can affect the ability of nature reserves to conserve medicinal plants by providing appropriate habitats for these species. Polley et al. and Šímová et al. have shown that precipitation patterns are the major driver of vegetative cover. It's not surprising that precipitation was the major force behind distribution of these species.

In general, this analysis allows us to evaluate the ability of nature reserves to protect medicinal plants by using current distribution maps and predicting potential distributions. Our study indicates that applying Maxent, followed by GIS is an effective strategy for establishing a map for flora species richness of medicinal plants on a large spatial scale. The map may then be used to inform the development of effective management plans for these endangered and valuable medicinal plants. Our results suggest that it is important for land managers to monitor changing trends in annual mean precipitation and precipitation seasonality in nature reserves because these two variables can determine the ability of the reserves to conserve medicinal plants.

The existing system of nature reserves with the highest medicinal plant species richness is also the areas we identified as areas with high ISRI values. These areas will continue to play important roles in medicinal plant preservation and should therefore be monitored over the long term to ensure protection of these and other species. This should be done in conjunction with increasing protection levels in these reserves, establishing protection districts, and performing in-situ and ex-situ conservation, especially for endangered plants.

We used ISRI to assess the ability of existing nature reserves to conserve medicinal plants because the nature reserves with high ISRI values have large areas with generally high plant species richness and are also suitable for the growth of medicinal plants. Thus, ISRI values can serve as a proxy for the level of conservation importance for each nature reserve. To conserve these species, the nature reserves must be monitored in terms of their annual mean precipitation and precipitation seasonality, namely, precipitation. These areas should be considered high priorities for acquiring additional reserve lands and conservationists working in these regions must adopt in-situ and ex-situ conservation methods and medicinal plant harvesting guidelines.

Conservation officials may be able to propose new conservation and observation areas due to the high ISRI values we recovered in some areas. For the areas with high plant species richness outside of existing nature reserves, it will be important to establish more areas of conservation and cultivation in order to retain and use the medicinal plants. Even so, protection and cultivation measures should not be hastily implemented and it will be important to commit in reaching short-term conservation goals. In addition, long-term planning for the conservation of these plants can be attained through the use of analyses of flora species richness, such as the ones we describe here.

Finally, our study provides new information on the distribution of medicinal plants in northeast China. Our research strategy can be used to conserve medicinal plants by informing effective and efficient management planning in the nature reserve system. Software tools to model species distributions are becoming increasingly sophisticated and will improve our ability to accurately predict the impact of climate change, known to be an important factor in species survival. Further work needs to be done to study the impact of climate change on medicinal plant species distributions. Specifically, the future potential distribution can be modeled under a variety of climatic scenarios and areas with appropriate habitats can be identified as important reserves and assisted migration can be used to transport plants to those areas. Climate change will need to be taken into consideration in future analyses to further improve accuracy of species distribution predictions, and ultimately inform the planning of reserve areas to protect flora species richness providing solid data to support and lead the development of medicinal plant protection in the future. Overall, the results of our study can be used not only for protecting medicinal plants, but also for general plant management, ecological construction, and geographical surveying in different regions.

Conclusions

Northeast China is rich in medicinal plant resources, but the region is affected by human disturbances and environment factors. Our study provides new verification of the richness of medicinal plants in northeast China by measuring ISRI values. This research is of use to study the effects of environmental factors on population growth and the distribution of medicinal plant species. We think that the areas of nature reserves are much smaller than the areas with high flora species richness. Nature reserves should be established to realize short-term conservation goals and after consideration of long-term projections of plant species richness and climate change, especially the change of precipitation.

https://www.sciencedirect.com/science/article/pii/S1574954114000776

Vocabulary

disturbance 英 [dɪˈstɜːbəns] 美 [dɪˈstɜːrbəns] n. 困扰；打扰；变乱
geographical 英 [ˌdʒiːəˈɡræfɪkl] 美 [ˌdʒiːəˈɡræfɪkl] adj. 地理学的；地理的
anecdotal 英 [ˌænɪkˈdəʊtl] 美 [ˌænɪkˈdoʊtl] adj. 轶事的；趣闻的
accurate 英 [ˈækjərət] 美 [ˈækjərɪt] adj. 精确的；准确的
variable 英 [ˈveəriəbl] 美 [ˈveriəbl] n. 可变因素；变量
seasonality 英 [ˌsiːzəˈnæləti] 美 [ˌsiːzəˈnæləti] n. 季节性
proxy 英 [ˈprɒksi] 美 [ˈprɑːksi] n. 代理人；代替物

Useful Expressions

1. in conjunction with   与……协力

The army should have operated in conjunction with the fleet to raid the enemy's coast.

陆军本应与舰队共同行动，突袭敌方海岸。

2. commit 英 [kə'mɪt] 美 [kə'mɪt] vt. 犯罪
I have never committed any crime.
我从来没犯过罪。

Questions

1. What are the most important variables for predicting the distribution of these 49 species according to the passage?

2. Why is ISRI used to assess the ability of existing nature reserves to conserve medicinal plants?

3. What is the significance of this study?

2.3 General Situation of Medicinal Plants Resources in Administrative Regions of China

Passage 1

Plant Resources of Sichuan (Part 1)

The climate of Sichuan Province can be classified as wet subtropic. As the geological conditions are complicated, many kinds of environmental conditions are existing, thus they house very abundant botanical resources. Some 10,000 species of higher plants are recorded (equalling one third of the national register); they belong to over 1,600 genera in 230 families. Nationwide Sichuan Province's biodiversity holds the second place after Yunnan in ferns (670 spp.) and angiosperms (8,453 spp.), and in gymnosperms (88 spp.). Sichuan's agriculture has a long history, so up to now it has over 1,500 cultivated species, most of them original, so they are well adapted to environment and bear a great potential for development.

Some General Characteristics of the Sichuanese Flora

1) Tropic-Subtropic and Temperate Species Dominate

Sichuan lies on the border of subtropics and tropics, its landscape is notably diverse, which has proved beneficial for a rich evolution. According to not yet complete investigations using the Engler system, Sichuan bears 9,254 species, 1,621 genera and 232 families of vascular plants (whereof ferns: 708 species, 120 genera, 41 families; gymnosperms: 88 species, 27 genera, 9 families; angiosperms: 8,453 species, 1,474 genera, 182 families). According to their distribution, the species can be classified as tropical (1.7%), both tropical and subtropical (63.2%), exclusively subtropical (mere 0.9%), temperate (29.2%) and arctic (5.6%). The distribution of genera is 2.9% for tropical, 55.9% for tropical-subtropical, 4.5% for exclusively subtropical, 34.5% for temperate and 2.8% for arctic areas. From these figures it is evident that Sichuan's flora is largely characterized by tropical-subtropical plants. Nonetheless, the high mountain areas in Sichuan's west and southwest also provide important centers of diversity for temperate genera such as Rhododendron L., Primula L., Gentiana L. and Saussurea L. Arctic and tropic plants contribute only a minor part to Sichuan's flora.

2) Endemic Genera Are Numerous

Sichuan's area is large and its ecological conditions vary from alpine-arctic to subtropical. The recent rising of the Himalayas and the relatively mild glacial periods also were auspicious for the formation of an abundant biodiversity. For example, half a dozen species are considered sufficiently diversified to form single-species families as Cercidiphyllum japonicum var. sinense, Davidia involucrata, Tetracentron sinense, Sargentodoxa cuneata and Eucommia ulmoides-all endemic to the mixed forests in Sichuan's southwest. Furthermore, at low altitude forests or cultivated near temples Gingko biloba can often be found. Single-species genera are even more numerous (29). We list a few endemic to south-west China: Tapiscia Oliv., Dichotomanthes Kurz., Emmenopterys Oliv., Davidia Baill., Fargesia Franch., Spenceria Trimen, Salweenia E. G. Baker, Kingdonia Balf. F. et W. W. Sm., Psilopeganum Hemsl., Atropanthe Pascher, Dickinsia Franch., Dinolimprichtia Wolff, Siphocranion Kudo, Sinojohnstonia Hu, Sinofranchetia Hemsl., Psammosilene W. C. Chen, Vertrilla Franch., Itoa Hemsl., Hosiea Hemsl. et Wils., Cathaya Chun et Kuang, Metasequoia Hu et Cheng, Heterolamium C. Y. Wu. Likewise, oligo-species genera often are relics of very distinct taxa, too some Sichuanese examples are Faberia Hemsl., Pterygiella Oliv., Speranskia Baill., Bretschneidera Hemsl., Dipteronia Oliv., Chimonanthus Lindl., Biondia Schltr., Asteropyrum J. R. Drumm et J. Hutch., Dipelta Maxim., Przewalskia Maxim., Notopterygium Boiss., Urophysa Ulbr., Oligobotrya Bak., Ancyclostemon Craib., Isometrum Craib., Dipoma Franch., Hemilophia Franch., Diuranthera Hemsl., Dysosma Woodson, Nannoglottis Maxim., Xanthopappus Winkler., Melliodendron Hand.-Mazz., Sindechites Oliv., Meehania Britton., Hanceola Kudo, Ostryopsis Decne., Rostrinucula Kudo, Bolbostemma Franquet, Gymnotheca Decne., Thyrocarpus Hance, Schnabelia Hand.-Mazz., Tetrapanax K. Koch, Torricellia DC., Solms-Laubachia Muschler, Sinojackia Hu, Discocleidium Pax ex Hoffm., Hemiboea C. B. Clarke, Tremacron Craib., Corallodiscus Batalin, Pararuellia Bremek. and Kinostemon Kudo. These great numbers of indigenous plants indicate Sichuan's flora is both ancient and special: among the angiosperms 464 species (5.48%) are currently considered as true endemics.

3) Many Relict Species Prove Antiquity

Sichuan's vegetational history is quite long. Its western part has been above sea level since Paleozoic times. As elsewhere, during Mesozoic times it had cycadophyte-pteridophyte covering. The climate became dryer during the Cretaceous (most forests disappeared), but in the Tertiary became warmer again. Most crucial, East Asia was lucky to pass a milder Quarternary glacial period than comparable areas in Europe or North America, thus many species could survive here. Although during the quartery glacial periods in the west Sichuan high mountain ridges glaciers were formed, which has not affected the survival of many ancient species, many of them moving south via the Hengduan Mountains (i. e. fir forests moved down to 1,500 m in Xichang, or even 1,100 m in Panzhihua), and then radiating again in the Quarternary. Ancient plants found in present Sichuan comprise ferns such as the paleozoic Psilotum, Angiopteris, the mesozoic Osmunda, Dicraepteris, Hieropteris, the jurassic Cyathea and Cibofinum, the cretacreous Plagiogyria and the tertiary Pteris, Lycopodium and Lygodium. Sichuan harbors three very famous "living fossils": Gingko, Cathaya and Metasequoia. Other ancient gymnosperms are the prejurassic Cycas, the cretaceous Pinus, Picea, Torreya, Cephalotaxus, as well as Keteleeria, Abies, Tsuga, Cryptomeria, Cunninghamia, Podocarpus and Ephedra in the Tertiary. There are about 50 ancient angiosperm families formed in the Cretaceous as the Trochodendraceae, Cercidiphyllaceae, Moraceae, Celastraceae, Rhamnaceae, Aceraceae, Magnoliaceae, Lauraceae, Ranunculaceae, Fagaceae, Hamamelidaceae, Betulaceae, Juglandaceae, Menispermaceae, Ericaceae et al., as well as 30 Tertiary families like the Myricaceae, Simarubaceae, Saururaceae, Alangiaceae, Hippocastanaceae, Nyssaceae, Sabiaceae, Flacourtiaceae, Stachyuraceae, Theaceae, Bretschneideraceae, Styraceae, Verbenaceae, Davidiaceae, Symplocaceae, Bignoniaceae, Polygalaceae, Meliaceae, Eleagnaceae, Calycanthaceae and Eucommiaceae.

4) Gymnosperm Diversity

With 88 species in 27 genera and 9 families Sichuan is China's richest province in gymnosperms (nationwide we count 232 species, 34 genera, 10 families), so 79.4% of the Chinese genera or 37.9% of China's species are represented. Most abundant are the Pinaceae (43 spp.), followed by the Cupressaceae (17 spp.), Cephalotaxaceae (6 spp.), Taxodiaceae, Podocarpaceae, Ephedraceae (5 spp. each), Gingkoaceae and Cycadaceae (1 sp. each). 14 species (15.9%) are endemic. The other gymnosperms are occuring in neighboring provinces, too, as in Yunnan (46 spp.), Hubei (31 spp.), Gansu and Guizhou (each 30 spp.), Tibet and Shaanxi (each 23 spp.) and Qinghai (only 9 common spp.).

The plant community of Sichuan developed from a paleo-mediterranean-cretaceous plant community and can now be considered as subtropical. For its rich evolution the rising of the western part of the province was very auspicious: Many habitats become disconnected during glacial periods providing many oppurtinities for further diversification. The differences between the alpine and subtropic areas can be summarized as follows: In the western mountains, albeit the number of genera being low (mainly Abies, Picea,

Pinus, Tsuga, Larix and Sabina), their distributions are wide, but according to geomorphology, mosaic and gradient distributions do often form. Here Picea and Abies are particulary dominant, with the exception of Abies fargesii West Sichuan contains all provincial representatives of these two genera, and is being a major center of diversity nationwide. Here you'll find extremely cold-resistant species like Picea likiangensis var. balfouriana and Abies squamata which survive up to 4,500 m and 4,600 m respectively. On the other hand, Abies ernestii and Picea wilsonii are more adapted to warmer environments at 2,200—2,800 m. Although Abies ernestii belongs to the genus, it grows well on dry slopes, whereas Picea wilsonii grows in damp ravines. Many examples like this demonstrate a high degree of diversification.

On the other hand, in the eastern part of the province, gymnosperms do not coin the whole vegetation, but many tertiary relict plants survive. Interesting examples are Metasequoia glyptostroides, Glyptostrobus pensilis, Cathaya argyrophylla, Fokienia hodginsii, Amentotaxus argentotaenia, Thuja sutchuenensis, Gingko biloba, Pseudotsuga sinensis, Keteleeria davidiana, Cunninghamia lanceolata, Torreya fargesii, Taxus sinensis, Cephalotaxus fortunei, etc.

http://www.blasum.net/holger/wri/biol/sichuanp.html

Vocabulary

abundant 英 [ə'bʌndənt] 美 [ə'bʌndənt] *adj.* 大量的
angiosperm 英 [ˌændʒɪəˌspɜːm] 美 [ændʒɪrˌspɜːm] *n.* 被子植物
exclusively 英 [ɪk'skluːsɪvlɪ] 美 [ɪk'sklusɪvlɪ] *adv.* 唯一地
arctic 英 ['ɑːktɪk] 美 ['ɑrktɪk, 'ɑrtɪk] *n.* 北极圈；北极地带
glacial 英 ['ɡleɪʃl] 美 ['ɡleʃəl] *adj.* [地]冰的；冰河期的
auspicious 英 [ɔː'spɪʃəs] 美 [ɔ'spɪʃəs] *adj.* 吉利的；有希望的；有利的
Cercidiphyllum japonicum var. sinense   日本连翘
Davidia involucrate   珙桐
Tetracentron sinense   水青树
Sargentodoxa cuneata   大血藤；红藤
alpine 英 ['ælpaɪn] 美 ['ælˌpaɪn] *adj.* 高山的
Eucommia ulmoides   杜仲
relict species   [植]孑遗种
antiquity 英 [æn'tɪkwəti] 美 [æn'tɪkwɪti] *n.* 古人；古代；古物
paleozoic times   古生代
mesozoic times   中生代
cycadophyte 英 [saɪ'kædəfaɪt] 美 [saɪ'kædəfaɪt] *n.* 苏铁亚纲植物
pteridophyte 英 ['terɪdəʊfaɪt] 美 ['terədoʊˌfaɪt] *n.* 羊齿类
gymnosperm 英 ['dʒɪmnəspɜːm] 美 ['dʒɪmnəˌspɜːm] *n.* 裸子植物

provincial 英 [prəˈvɪnʃl] 美 [prəˈvɪnʃəl] *adj.* 省的；州的；地方的
ravine 英 [rəˈviːn] 美 [rəˈvin] *n.* 沟壑；深谷

Useful Expressions

1. thus 英 [ðʌs] 美 [ðʌs] *adv.* 于是；因此
Neither of them thought of turning on the lunch-time news. Thus Caroline didn't hear of John's death until Peter telephoned.
两个人谁也没有想到打开电视看午间新闻，所以直到彼得打来电话，卡罗琳才得知约翰的死讯。

2. house 英 [haʊs] 美 [haʊs] *vt.* 给……提供住房；收藏；安置
Their villas housed army officers now.
他们的别墅里现在住着军官。

3. disconnect 英 [ˌdɪskəˈnekt] 美 [ˌdɪskəˈnɛkt] *vt.* 断开
The device automatically disconnects the ignition when the engine is switched off.
当引擎关掉时，这个装置会自动切断点火开关。

4. albeit 英 [ɔːlˈbiːɪt] 美 [ɔlˈbiɪt, æl-] *conj.* 虽然；即使
Charles's letter was indeed published, albeit in a somewhat abbreviated form.
尽管有所删节，但查尔斯的信确实被刊登出来了。

5. coin 英 [kɔɪn] 美 [kɔɪn] *vt.* 制造硬币；杜撰；创造
Jaron Lanier coined the term "virtual reality" and pioneered its early development.
杰伦·拉尼尔创造了"虚拟现实"这个词，并成为推动其早期发展的先驱人物。

Questions

1. How abundant are the botanical resources in Sichuan?
2. Why the plant community of Sichuan is considered as subtropical?
3. What are the differences between the alpine and subtropic areas?

Passage 2

Plant Resources of Sichuan (Part 2)

Phytogeography of Sichuan

A glance on the map of present distribution of tree societies shows that Sichuan can be divided into three major landforms. In the eastern part lies the subtropic Sichuan basin (including the major cities as Chengdu, Chongqing, etc.) surrounded by mostly lower mountains (1,500—2,000 m), such as the Micang and Daba Mountains in the north, the Wu and Qiyao Mountains in the east, the Dalou Mountains in the south, Longmen, Emei, Lower and Upper Liang Mountains in the west. The north-western part of the province, however, is characterized by the adjacent Qinghai-Tibet plateau grasslands (3,500—4,500 m). Between the basin and the grassland a zone of high mountain coniferous forests can be found (3,000—4,000 m).

1) The Basin (East & Central Sichuan) and Evergreen Broad-leaved Woodlands (South-west Sichuan)

This region is situated south east of the counties Pingwu, Maowen, Baoxing, Kangding, Luding, Jiulong and Muli, it thus comprises nearly the whole province with the only exception of the Gansun and Aba Tibetan autonomous regions. In the Sichuan basin and its surrounding lower mountains the climate is hot in summer, mild in winter; frosts are seldom; rain is plenty (1,000—1,200 mm p. a.) all over the year; it's moist and misty. Average annual temperatures mainly range between 16—19℃. In the Hengduan mountains regions in south-west Sichuan, monsoon periods gradually become more distinct: Winters are quite dry, whereas summers are very damp. The typical vegetation comprises subtropical evergreen broad-leaved forests (Fagaceae, Lauraceae, Theaceae, Magnoliaceae, Symplocaceae), subtropical coniferous woods (Pinaceae, Taxodiaceae, Cupressaceae) and bamboo stands. The natural conditions are very good in this area, but as the densely populated basin is mainly used for agriculture, and wild plants are confined to wastelands, riversides or village woods. The east is characterized by south-eastern monsoons; the climate is very humid; frosts are seldom; there is no distinguished arid period, which supports humid evergreen broad-leaved forests (Lauraceae, Fagaceae, Theaceae) and subtropical coniferous forest (mainly composed of Pinus massoniana, Cunninghamia lanceolata and Cupressus funebris). The fertile center of the basin is dominated by a long agricultural history, so dominant trees are Pinus, Abies, Cupressus, many Bambusoidae, Lauraceae, Alnus, Eucalyptus, etc. Citrus sinensis and related species are widely cultivated. In these regions, two to three harvests a year are possible, so agricultural crops dominate, but an estimated 1,000 to 1,500 species of medicinal herbs still can be found. Representative families are Moraceae, Ranunculaceae, Crucifera, Leguminosae, Rutaceae, Meliaceae, Combretaceae, Umbelliferae, Scrophulariaceae, Lamiaceae, Campanulaceae, Compositae, Alismataceae, Gramineae, Liliaceae and Zingiberaceae. The vegetation in the surrounding lower mountains (1,500 to 2,000 spp. of medicinal plants) is far more heterogenous, as it is largely influenced by latitude and precipitation gradients. In the south and southwest, temperatures are high; precipation is abundant; in the evergreen broad-leaved forests Schima, Gordonia and many liana shrubs are typical. In the west and northwest of the basin although it is slightly colder than in the south, precipitation is still high. So in these broad-leaved forests Cinnamomum wilsonii dominates, but also liana shrubs, all kind of epiphytic and parasitic plants as well as mosses are common. On the colder slopes of the Northeast Sichuanese Daba Mountains as also in the mountains east of the basin, the forests begin to be dominated by Fagaceae, Theaceae (yet Lauraceae are still quite numerous).

Summarizing, plants found in the mountains around the basin are Auricularia auricula, Tremella fuciformis, Polyporos umbellatus, Lygodium japonicum, Equisetum hiemale, Cibotium barometz, Osmunda japonica, Cyrtomium fortunei, Davallia mariesii, Pyrrosia lingua, etc., for the fungi and ferns; Pinus, Taxus, Cupressus, Ephedra as representative gymnosperm genera; common angiosperm families are Moraceae, Urticaceae, Loranthaceae, Aristolochiaceae, Polygonaceae, Amaranthaceae, Phytolaccaceae, Caryophyllaceae, Ranunculaceae, Lardizabalaceae, Berberidaceae, Menispermaceae, Magnoliaceae, Lauraceae, Papaveraceae, Cruciferae, Saxifragaceae, Eucommiaceae, Rosaceae, Leguminosae, Geraniaceae, Rutaceae, Meliaceae, Euphorbiaceae, Coriaceae, Rhamnaceae, Vitaceae, Theaceae, Araliaceae, Umbelliferae, Cornaceae, Ericaceae, Oleaceae, Gentianaceae, Asclepiadaceae, Convolvulariaceae, Lamiaceae, Solanaceae, Acanthaceae, Plantaginaceae, Rubiaceae, Valerianaceae, Caprifoliaceae, Dipsacaceae, Cucurbitaceae, Campanulaceae, Compositae, Gramineae, Cyperaceae, Araceae, Stemonaceae, Liliaceae, Dioscoreaceae, Iridaceae, Zingiberaceae, Orchidaceae, etc.

Of the western Sichuan highlands, the southern part is also influenced by the monsoon climate in the procumbent lower mountains surrounding the basin, drought and rain period are clearly distinguishable. This area is characterized by broad-leaved woods formed by drought-resistant species like Fagaceae or subtropical conferous forests (Pinus yunnanensis, Keteleeria evelyniana, Cupressus duclouxiana). About 2,500 species of medicinal plants are located here, such as Polyporus, many ferns, Ranunculaceae, Lardizabalaceae, Phytolaccaceae, Berberidaceae, Menispermaceae, Magnoliaceae, Rosaceae, Leguminosae, Araliaceae, Umbelliferae, Ericaceae, Gentianaceae, Scrophulariaceae, Dipsacaceae, Campanulaceae, Compositae, Gramineae, Araceae, Stemonaceae, Liliaceae, Dioscoreaceae and Iridaceae. At higher altitudes (1,000 to 2,000 m) in the same evergreen forests other species become abundant, such as Cinnamomum wilsonii, Lindera glauca, Phellodendron chinense, Mahonia gracilipes, Schefflera octophylla, Sch. delavayi, Helwigia japonica, Hedera nepalensis var. sinensis, Akebia trifoliata var. australis, Sinomenium acutum, Cocculus trilobus, Paris polyphylla, Liriope spicata, Cibotium barometz, Cyrtomium fortunei, Pyrrosia petiolosa and Lepisorus thunbergianus.

Where evergreen and deciduous broad-leaved forest mix (2,000—2,600 m) plant societies change again, notable representatives of this area are Ilex cornuta, Eucommia ulmoides, Gastrodia elata, Coptis chinensis, Cimicifuga foetida, Asarum sieboldii, Panax bipinnatifidus, Anthriscus sylvestris, Convallaria majalis, Veratrum nigrum, Hemsleya amabilis, Pyrola rotundifolia, Astilbe chinensis, Polygonum runcinatum, Dysosma veitchii, Oxalis griffithii, Actinidia spp., Aralia chinensis, Celastrus orbiculatus, etc.

In the southwest Sichuan river valleys at low altitudes some wild tropical plants are found: Oroxylum indicum, Calotropis gigantea, Opuntia dillenii. Others are cultivated like Andrographis paniculata, Pogostemon cablin and Amomum villosum. In subalpine shrubland Iphigenia indica is common.

2) Alpine Coniferous Forests (Western Sichuan)

This region is located west of the counties Pingwu, Maowen, Baoxing, Kangding, Jiulong and Muli and southeast of Dengke, Zhuqing (Dege County), Hexi (Seda County), Nanmuda and Chazhenliangzi (Rangtang County), Huangshengguan (Songpan County) and Ruoergai. It comprises the whole counties of Nanping, Songpan, Heishui, Lixian, Maerkang, Jinchuan, Xiaojin, Danba, Kangding, Daofu, Qianning, Xinlong, Dege, Baiyu, Yajiang, Yilun, Batang, Xiangcheng, Daocheng, Derong, Litang and parts of the counties Ruoergai, Hongyuan, Rangtang, Wenchuan, Maowen, Gansun, Luhuo, Dengke and Muli. This zone borders to the Qinghai-Tibet high plateau, and its moutains are steep. While being rich in geothermic energy, its geomorphology is complex. Its eastern mountain valleys are abundant in spruces, firs and pines, forming the (somewhat overlogged) major wood resource for Sichuan. In its west, subalpine shrublands and meadows begin to dominate. For some river valleys, an average annual precipitation of ca. 600 mm and an annual medium temperature of 12 ℃ might be given, but local climatic conditions are varying very much according to altitude.

The subalpine shrublands and meadows are rich in medicinal herbs, such as Fritillaria cirrhosa, Rheum officinale, Gentiana macrophylla, Astragalus membranaceus, A. folridus, A. chrysopteris, A. tongolensis, Paeonia veitchii, P. suffruticosa, Rosa banksiae, Lamiophlomis totata, Dracocephalum tanguticum, Meconopsis spp., Aconitum spp., Berberis spp., Gentiana spp., Lloydia serotina, L. tibetica, Arisodus luridus, Stellera chamaejasme, Heracleum candicans, Thlaspi arvense and Descurainia sophia. Among major aromatic oil plants are Rhododendron cephalanthum, Rh. fastigiatum, Rh. flavidium and Sabina spp. In alpine rock deserts above 4,500 m medicinal herbs are Fritillaria cirrhosa, Saussurea spp., Soroseris spp., Eriophyton wallichii, Lagotis ramalana, L. integra, L. brevituba and Solms-Laubachia pulcherrima.

3) High Plateau Grass and Shrublands (North-Western Sichuan)

This region is situated north-west of the line Dengke, Zhuqing (Dege County), Hexi (Seda County), Nanmuda (Rangtang County), Aba, Chazhenliangzi, Huangshengguan (Songpan County) to Ruoergai. It wholly or partially contains the counties Shiju, Aba, Dengke, Dege, Seda, Gansun, Luhuo, Tangrang, Hongyuan, Ruoergai and Songpan and belongs to the Qinghai-Tibet high plateau. From 3,300—4,000 m in the east, it gradually rises to 3,900—4,500 m in the west. The river valleys are broad and not deeply incised and hills are quite flat, so mostly the height difference between river bed and surrounding hilltops does not exceed 400 m. This climate is cold (medium temperature ca. -1 ℃) and dry (precipitation about 650 mm); irradiation is intensive. Grasslands (Roegneria nutans, Poa pratensis, Agrostis schneideri, used for grazing yaks and sheep) dominate, intertwined with shrublands and marshlands. At some places, fragmented subalpine coniferous woods occur.

Medicinal plants populating shrub and grasslands are Fritillaria cirrhosa, Rheum officinale, Lasiosphaera nipponica, Thamnolia vermicularia, T. subvermicularia, Nardostachys chinensis, Cirsium souliei, Aconitum spp., Saussurea spp., Lagotis spp., Meconopsis spp., Przewalskia tangutica. Other plants are rich in starch, such as Polygonum viviparum, P. sphaerostachyum and Potentilla anserina. Last but not least, in the grassland of Shiju County edible fungi (as Tricholoma sordidum) are numerous.

In short, the vertical distribution is quite distinct anywhere in Sichuan: We can distinguish the agricultural region (below 1,200 m), a surrounding secondary shrubland area (900—1,500 m), an evergreen broad-leaved forest belt (1,400—2,000 m), a zone where evergreen broad-leaved and deciduous forests mix (1,800—2,500 m), subalpine and alpine conifers (2,500—3,200 m), alpine shrublands and meadows (3,200—4,500 m) and alpine rock vegetation (above 4,500 m).

http://www.blasum.net/holger/wri/biol/sichuanp.html

Vocabulary

phytogeography 英 [ˌfaɪtəʊdʒɪˈɒɡrəfɪ] 美 [ˌfaɪtoʊdʒiːˈɒɡrəfiː] *n.* 植物地理学
adjacent 英 [əˈdʒeɪsnt] 美 [əˈdʒesənt] *adj.* 毗邻的；(时间上) 紧接着的
coniferous 英 [kəˈnɪfərəs] 美 [kəˈnɪfərəs] *adj.* 松类的；结球果的
wasteland 英 [ˈweɪstlænd] 美 [ˈwestˌlænd] *n.* 荒地；荒漠；贫乏
humid 英 [ˈhjuːmɪd] 美 [ˈhjumɪd] *adj.* 潮湿的；湿润的
frosts 英 [frɒst] 美 [frɔːst] *n.* 霜冻；结霜；严寒；寒冷
arid period 旱季
heterogenous 英 [ˌhetəˈrɒdʒənəs] 美 [ˌhetəˈrɒdʒənəs] *adj.* 异种的；异质的
gradient 英 [ˈɡreɪdiənt] 美 [ˈɡreɪdiənt] *adj.* 倾斜的
liana shrub 藤本灌木
meadow 英 [ˈmedəʊ] 美 [ˈmedoʊ] *n.* 草地；牧场
medium 英 [ˈmiːdiəm] 美 [ˈmidiəm] *n.* [生]培养基；培养液
plateau 英 [ˈplætəʊ] 美 [plæˈtoʊ] *n.* 高原
marshland 英 [ˈmɑːʃlænd] 美 [ˈmɑːrʃlænd] *n.* 沼泽地；沼地
fragment 英 [ˈfræɡmənt] 美 [ˈfræɡmənt] *vt.* 破裂；分裂
edible 英 [ˈedəbl] 美 [ˈɛdəbəl] *adj.* 可以吃的；可食用的
fungi 英 [ˈfʌŋɡiː] 美 [ˈfʌndʒaɪ] *n.* (fungus 的复数) 真菌

Useful Expressions

1. be confined to   限于；禁闭
The woman will be confined to a mental institution.

这个女人将被关进精神病院。
2. intertwine 英 [ˌɪntəˈtwaɪn] 美 [ˌɪntərˈtwaɪn] vt. 缠结在一起；使缠结
Their destinies are intertwined.
他们的命运交织在一起。

Questions

1. What will you find when looking at the map of present distribution of tree societies?
2. What plants are there in the western Sichuan highlands?
3. What plants are there in north-western Sichuan?

2.4 Regionalization of Chinese Medicinal Plants Resources

Passage 1

Regionalization of Chinese Material Medical Quality Based on Maximum Entropy Model: A Case Study of Atractylodes Lancea (Part 1)

Introduction

Regionalization, which means the division of regions, is a way that people extract the characteristic information of space and then classify and merge regions according to certain purposes in understanding the natural and social environment. The regionalization of Chinese material medical resources was to study Chinese material medical resources and their spatial variance in region system and regionalize the herbal medicine resources according to the similarity and variation of ecological characteristics. Regionalization of Chinese herbal medicine has been studied since the 1990s, and the data came from the results of the Third National Census of Chinese Material Medicine Resource. Great achievements in traditional Chinese medicine regionalization have been gained in the latest 20 years. But there still exist problems. For instance, the representativeness of sample is not enough, meanwhile, the accuracy and extent of ecology factor data are low, which results in the process of regionalization that has always been affected by human's subjective factors. At the same time, in Chinese material medical planting industry, there often exists the problem that although the index components of Chinese material medical content conforms to the standard of *Pharmacopeia*, they are not as good as the original regions of traditional Chinese medicine, even some of them cannot reach to the standard of *Pharmacopedia*. Exploring the relationship between ecological factors of Chinese herbal medicine production areas and medical quality, optimizing the planting areas of traditional Chinese medicine, solving the practical problems during the process of traditional Chinese medicine production, and improving the reliability and practicability of traditional Chinese medicine regionalization are the aims of the Chinese material medical quality regionalization.

Maximum Entropy Model is a species distribution model based on machine learning and the basics of ecology. It contains many advantages: The different types of ecological factors are compatible, and in the progress of calculation, the effects of human's subjective factors are eliminated and the model solely relies on the results of objective sampling. The model is widely used in studies of the coverage of animals and plants in terrestrial and marine ecosystems for its high accuracy. The distribution and quality of traditional Chinese medicine can be affected by natural environment. Habitat suitability is used to describe the quantitative index of relationship between ecological environment and the distribution of medical materials. We can establish the quantitative relationships between the ecological environment and the quality of traditional Chinese medicine by means of coupling correlation analysis, principle components analysis and clustering analysis.

As an important resource of medicinal plant, atractylodes is collected in national pharmacopoeia by China, Japan and Korea. Atractylodes has long been used as herbal medicine, and has significant features in its trueborn quality and geographical distribution. In this research, we used atractylodes as a case to study the regionalization method of Chinese material medical quality. Specifically, based on the complete sample data and using the habitat suitability as the linkage, we studied the relationships between the quality of medical materials and ecological environment by the Maximum Entropy Model and other statistical analysis techniques. We also used the similarity of environment to grade the quality of medical materials and at last we achieved a map of regionalization of Chinese material medical quality.

The relationship between biology and environment is a core issue in ecology. The distributional suitability of organisms in the environment can be evaluated by species distribution model. The species distribution (it is also called Habitat Suitability Model or Environment Niche Model) covers many different models. It mainly relies on species distribution data which is already known to us and a series of environment variables to explore the species niche and its potential distribution. According to the data of model needed and its own principle, it can be divided into three different kinds: Group Discriminant Model, Frame Model and Mechanism Model.

Entropy is an elementary concept in information theory. Shannon put forward the concept of entropy in an article of information theory domain in 1948 that entropy was used to describe the disorder of object, which was positively related to the degree of entropy, and the degree of uncertainness about object was called entropy. Principle of maximum entropy (PME) adopts the principle to treat all the known and unknown objects: PME admits the known things while it does not make hypotheses for unknown things, which means that it keeps the uncertainness of object and lower the crisis to satisfy all known information. Under those conditions, attaining a probability distribution with less subjective prejudice is a most reasonable distribution. The basic idea of PME is offering a certain training sample and choosing a model according with the training sample. Maximum Entropy Model should choose a probability distribution conform to the discovery, and model gives a uniform probability distribution to other situations. The theory of PME is expressed in the bio-ecology that one species would spread and extend to the greatest extent possiblilities without any constraint condition, and close to the uniform probability distribution. Steven Phillips and other people used the Java language to compile Maxent software on the basis of Maximum Entropy.

In the software of Maxent, define π as a probability distribution of finite set X. For each x in the finite set X, $\pi(x)$ as a probability distribution of this point and it must be non-negative. The sum of all $\pi(x)$ is up to 1. What we estimate is also a probability distribution. So we can get a formula to work out the entropy.

https://www.nature.com/articles/srep42417/

Vocabulary

regionalization 英 [ˌriːdʒənlɪˈzeɪʃən] 美 [ˌriːdʒənəlɪˈzeɪʃən] *n.* 分成地区
variance 英 [ˈveəriəns] 美 [ˈveriəns] *n.* 变化；变动
index 英 [ˈɪndeks] 美 [ˈɪnˌdɛks] *n.* 索引；[数]指数；指示；标志
pharmacopeia 英 [ˌfɑːməkəʊˈpeɪə] 美 [ˌfɑːməkoʊˈpeɪə] *n.* 处方汇编；药典
compatible 英 [kəmˈpætəbl] 美 [kəmˈpætəbəl] *adj.* [生] 亲和的
quantitative 英 [ˈkwɒntɪtətɪv] 美 [ˈkwɑːntəteɪtɪv] *adj.* 定量的；数量(上)的
clustering analysis 聚类分析；群集分析；集群分析；群聚分析；聚类分析法
Atractylodes 苍术属
entropy 英 [ˈentrəpi] 美 [ˈɛntrəpi] *n.* 熵；平均信息量；负熵
constraint 英 [kənˈstreɪnt] 美 [kənˈstrent] *n.* 约束；限制；强制

Useful Expressions

1. merge 英 [mɜːdʒ] 美 [mɜːrdʒ] *vt.* 融入；(使)混合；相融；渐渐消失在某物中
My life merged with his.
我和他的生活合而为一。
2. spatial 英 [ˈspeɪʃl] 美 [ˈspeʃəl] *adj.* 空间的；受空间条件限制的
His manual dexterity and fine spatial skills were wasted on routine tasks.
他灵巧的动手能力和杰出的空间识别能力都浪费在日常事务上了。
3. domain 英 [dəˈmeɪn] 美 [doʊˈmeɪn] *n.* 范围；领域
This information should be in the public domain.
这一消息应该为公众所知。

Questions

1. What is the purpose of regionalization of Chinese material medical resources?
2. What is entropy?
3. How can we evaluate the distributional suitability of organisms in the environment?

Passage 2

Regionalization of Chinese Material Medical Quality Based on Maximum Entropy Model: A Case Study of Atractylodes Lancea (Part 2)

Materials and Methods

Atractylodes has a long history of application and significant features in its trueborn quality, and is selected in this research. Atractylodes is collected from sampling sites all over the nation and at each experimental site we sample duplicated samples. After a unified and standardized test, we can gain the average content data of atractylodin in each sampling site. According to the stipulation of atractylodin content (content of atractylodin in the samples of oven dried atractylodes should not be less than 0.3%) from the standard of *Chinese Pharmacopeia*, sampling sites were divided into two parts: One is atractylodin with lowercontent than the standard of *Pharmacopeia*, and the other is atractylodin with higher content than the standard of *Pharmacopeia*. After that, we used Maximum Entropy Model to calculate the habitat suitability of atractylodes separately to obtain the particular characteristic of medicinal material in habitat in the case that atractylodin is higher than the standard of *Pharmacopeia* and atractylodin is lower than the standard of *Pharmacopeia*. Finally, by analyzing the promotive and inhibitive effect of habitat condition on accumulation of atractylodin, we obtained the quality regionalization map of atractylodes in national scale, and revealed the connections between the quality of medicinal material and habitat condition.

Data of Ecological Factors

Meteorological data came from the dataset of standard climatological annual and monthly value which was obtained from 722 Chinese ground meteorological observation stations from 1950 to 2000, and the dataset of climatological annual, monthly and daily value gaining from 752 Chinese ground meteorological observation stations and automated stations since 1951. By scientific analysis and calculation, we obtained the meteorological factor data for this study.

The data of soil type is arranged according to the 1:1,000,000 Scare Soil Map of China (1995) offered by the Second National Land Survey. This data can provide model input parameters for modeler. It can also be used to study eco-agricultural division, food security, climate change and so on in agricultural perspective.

Terrain data includes elevation, slope and aspect.

The data of vegetation type is arranged according to the sub-type vegetation data that come from to the 1:1,000,000 Scale Vegetation Map of China published by Institute of Botany, Chinese Academy of Sciences.

Comprehensive meteorological indices come from warmth index and coldness index in Kira index, and humidity index that was modified from Kira index by Xu Wenduo.

The Distribution Information of Atractylodes

The geographic range of atractylodes was further precisely checked by continuous field investigation from July to November in 2011. In each sampling site, altimeter was utilized to determine elevation, GPS was used to measure longitude and latitude, and the investigation of plant population was carried out. According to The Plant List (http://theplantlist.org), atractylodes lanceae was regarded the species name in this research. 2—5 samples were obtained from each of the 20 sample sites.

The Chemical Composition Content of Atractylodes

Gas Chromatography-Mass Spectrometer was used in chromatographic analysis of the volatile oil component standard substances in this study to get the standard curve of standard substances. The method is verified to be stable after repetitive experiments. And then, chromatographic analysis was operated with samples from the field. 4—5 replicate samples were obtained from each sample site. Comparisons of chromatogram between samples and standard substances were performed and finally 4 kinds of chemical composition content value of atractylodes were obtained.

Selection of Ecological Factors

The correlation coefficient between each ecological factor and atractylodin content was calculated by relevance strategy, and the data in descending order was sorted according to absolute value of correlation coefficient. We found that temperature played a key role in the accumulation of atractylodes essential oil component and precipitation had a significant influence on the survival of atractylodes by experience strategy and the living habits of atractylodes. With the consideration that climatological monthly value was better than annual value in reflecting the requirement of climate conditions by species, these ecological factors whose correlation coefficient with atractylodin was more than 0.2 were selected in this study.

The correlation coefficient of these factors and atractylodes content were calculated. We found that the mean precipitation in March and November, April and October whose correlation coefficient were more than 0.8, thus the mean precipitation in November and October were deleted. Finally, the correlation coefficient tree diagram of 17 ecological factors was obtained.

Maximum Entropy Model Calculation

Maxent Model needs two sets of data to operate, that is, (1) the distribution data of atractylodes in the form of latitude and longitude; (2) the ecological factor data of atractylodes in the study area. The operation processes of Maxtent model are as follows: First of all, obtain the distribution data and ecological factors of atractylodes in the study area; secondly, operate the model to choose suitable predictable ecological factors for the growth of atractylodes; thirdly, establish predictable model and predict the potential geographical distribution for target species; finally, work out a figure and evaluate the predictable accuracy of the model. Maxent assumes a uniform distribution of ecological factors at first and then proceeds iteration; the importance of every ecological factor in the formula is adjusted constantly during the process of iteration, which is called Training Gain. Maximizing the average probability of each grid value which is got in the end, thus the predictable distribution figure is obtained in the Maxent. Output value is continuous in a certain range, and each grid has a value to show the suitable degree of atractylodes to the ecological factor in the grid. The value in the grid net is a cumulative probability, and the bigger the grid value is, the more possible the species suitable degree becomes.

The parameter is set as follows when using Maxent 3.3.3 software: Convergence threshold is set at 0.00005; Maximum iterations is set at 1,000,000; other parameters are set to the default. 15% of parameters were selected as random test percentage randomly in samples, the rest data is set as the training data. The distribution data of atractylodes with higher and lower atractylodin content than the standard of *Pharmacopedia* was selected respectively to calculate the habitat suitability of the atractylodes in the national scale.

*https://www.nature.com/articles/srep*42417/

Vocabulary

trueborn 英 ['truːbɔːn] 美 ['truːˌbɔːn] *adj.* 嫡出的；纯正的；道地的
dataset   数据集
climatological 英 [ˌklaɪmətəˈlɒdʒɪkl] 美 [ˌklaɪmətəˈlɒdʒɪkl] *adj.* [气] 与气候学有关的
meteorological 英 [ˌmiːtɪərəˈlɒdʒɪkl] 美 [ˌmiːtɪərəˈlɒdʒɪkl] *adj.* 气象的
elevation 英 [ˌelɪˈveɪʃn] 美 [ˌɛləˈveʃən] *n.* 高处；高地；高度；海拔
indices 英 [ˈɪndɪsiːz] 美 [ˈɪndɪsiːz] (index 的复数) *n.* 指数
relevance strategy   关联策略
correlation coefficient   相关系数
convergence threshold   收敛阈值；收敛门限

Useful Expressions

1. promotive 英 [prəˈməʊtɪv] 美 [prəˈmoʊtɪv] *adj.* 促进性的
Preferential tax policy has promotive effect on attracting foreign investment.
优惠的税收政策对我国吸引外资起到了积极的促进作用。

2. inhibitive 英 [ɪnˈhɪbɪtɪv] 美 [ɪnˈhɪbɪtɪv] *adj.* 禁止的；抑制的

Objective: To study the inhibitive effect of tetrandrine to cell reaction on intraocular lens.

目的：研究汉防己甲素对人工晶状体表面细胞反应的抑制效果。

3. accumulation 英 [əˌkjuːmjəˈleɪʃn] 美 [əˌkjumjəˈleʃən] *n.* 积累；堆积物；累积量

The rate of accumulation decreases with time.

积累的速度随着时间变慢。

4. descending order   递减次序；降序排列

All the other ingredients, including water, have to be listed in descending order by weight.

所有其他原料，包括水在内，都需要根据重量依次降序排列。

5. verify 英 [ˈverɪfaɪ] 美 [ˈvɛrəˌfaɪ] *vt.* 核实；证明；判定

I verified the source from which I had that information.

我核实了我所获消息的来源。

Questions

1. Which method is atractylodes selected in this research?
2. What do you think is the most important chemical composition of atractylodes?
3. What does the correlation coefficient of the factors and atractylodes content indicate?

Chapter Three

Reserves of Medicinal Plants Resources

3.1 Concept of Medicinal Plants Resources Reserves

Passage 1

Conservation of Medicinal and Aromatic Plants in Brazil (Extract)

The Brazilian Vegetation

Approximately two thirds of the biological diversity of the world is found in tropical zones, mainly in developing countries. Brazil is considered the country with the greatest biodiversity on the planet, with nearly 55,000 native species distributed over six major biomes: Amazon (30,000), Cerrado (10,000), Caatinga (4,000), Atlantic rainforest (10,000), Pantanal (10,000) and the subtropical forest (3,000).

The Brazilian Amazon Forest (tropical rainforest) covers nearly 40% of all national territory, with about 20% legally preserved. This ecosystem is rather fragile, and its productivity and stability depend on the recycling of nutrients, whose efficiency is directly related to the biological diversity and the structural complexity of the forest. Giacometti estimates that there are about 800 plant species of economic or social value in the Amazon. Of these, 190 are fruit-bearing plants, 20 are oil plants, and there are hundreds of medicinal plants.

The "Cerrado" is the second largest ecological dominion of Brazil, where a continuous herbaceous stratum is joined to an arboreal stratum, with variable density of woody species. The cerrados cover a surface area of approximately 25% of Brazilian territory and around 220 species from cerrado are reported as used in the traditional medicine.

The "Caatinga" extends over areas of the states of the Brazilian Northeast and is characterized by the xerophitic vegetation typical of a semi-arid climate. The soils that are fertile, due to the nature of their original materials and the low level of rainfall, experience minor runoff. Various fruit species and medicinal plants have their centers of genetic diversity in this region, and the use of local folk medicines is common. Several important aromatic species are reported for this region, such as Lippia spp. and Vanillosmopsis arborea.

The Atlantic Forest extends over nearly the whole Brazilian coastline, and is one of the most endangered ecosystems of the world, with less than 10% of the original vegetation remaining. The climate is predominantly hot and tropical, and precipitation ranges from 1,000 to 1,750 mm. The land is composed of hills and coastal plains, accompanied by a mountain range. Several important medicinal species are found in this region, such as Mikania glomerata, Bauhinia forficata, Psychotria ipecacuanha, and Ocotea odorifera.

The territory of the Meridional Forests and Grasslands includes the mesophytic tropical forests, the subtropical forests, and the meridional grasslands of the states of southern Brazil. The climate is tropical and subtropical, humid, with some areas of temperate climate. The naturally fertile soils, associated with the mild climate, allowed a rapid colonization during the last century, mainly by European and, more recently, by Japanese immigrants. Several medicinal plants, such as chamomile (Matricaria recutita), calendula (Calendula officinalis), lemon balm (Melissa officinalis), rosemary (Rosmarinus officinalis), basil (Ocimum basilicum), and oregano (Origanum vulgare), were introduced and adapted by immigrants.

The Pantanal is a geologically lowered area filled with sediments which have settled in the basin of the Paraguay River. Pantanal flora is formed by species from both Cerrado and Amazon vegetation. More than 200 species useful for human and animal consumption as well as for industrial use have been recorded in this region.

Germplaam Conservation

In the last decade, serious efforts to collect and preserve the genetic variability of medicinal plants have been initiated in Brazil. The National Center for Genetic Resources and Biotechnology-Cenargen, in collaboration with other research centers of Embrapa (Brazilian Agricultural Research Corporation), and several universities, a program to establish germplasm banks for medicinal and aromatic species.

The first step is to establish criteria to define a species priority, based on economic and social importance, markets, and potential genetic erosion. Vieira and Skorupa proposed the following criteria to define priority (1) species with proven medicinal value including those containing known active substance (s) or precursor (s) used in the chemical-pharmaceutical industry with proven pharmacological action, or at least demonstrating pre-clinical and toxicological results; (2) species with ethnopharmacological information widely used in traditional medicine, and which are threatened or vulnerable to extinction; (3) species with chemotaxonomical affinity to botanical groups which produce specific natural products.

Conservation of threatened germplasm includes seed banks, field preservation, tissue culture, and cryopreservation. Seed storage is considered the ideal method; seeds considered orthodox can be dried and are able to be preserved at sub-zero temperatures (-20 ℃), while recalcitrant seeds, including most tropical species, lose their seed viability when subjected to the same conditions. Maintenance of the germplasm in field collections is costly, requires large areas, and can be affected by adverse environmental conditions. Tissue culture or cryopreservation techniques can be also considered in some cases.

The next step is to decide which germplasm conservation method will be applied: ex situ or in situ. In an ex situ procedure, the germplasm is collected from fields, markets, small farms, and other sites, in form of seeds, cuttings, underground systems, and sprouts. The collected samples should represent the original population with passport data and herbarium vouchers. In a long term, mutation can take place over the years in a cold chamber or in vitro conservation. In contrast, in situ conservation maintains population in its preserved natural area, allowing the evolutionary process to continue, although genetic reserves are subject to anthropogenic action and environmental effects. Most in situ conservation has focused in forest species, with some medicinal species included, such as Pilocarpus microphyllus and Aniba roseodora. The establishment of genetic reserves in Brazil has relied on national parks and conservation areas established by the environmental protection agency of Ibama, Brazil.

There are now five forest genetic reserves in Brazil: one in the Amazon Tropical Rainforest, State of Para; one in the Caatinga, State of Minas Gerais; two in the Cerrado in the Federal District; and one in the Meridional Forest (Subtropical) in the State of Santa Catarina. Four other genetic reserves are being created: two in the Atlantic Forest in the states of Rio de Janeiro and Espirito Santo, one in the Caatinga in the State of Piaui, and another in the Tropical Humid Forest in transition with Cerrado in the State of Minas Gerais. These reserves aim to conserve the most endangered species and those of greatest economic interest, including medicinal and aromatic plants.

The Brazilian program on medicinal germplasm conservation has three foci: (1) ethnobotanical studies; (2) germplasm collection and characterization; and (3) in situ conservation. Ethnobotanic and phytogeographic studies on the medicinal flora of Cerrado have been able to identify and collect genetic material for conservation. About 110 species used in traditional medicine were reported in the Cerrado region. Bibliography review and a herbaria search were carried out allowing an estimation of the medicinal potential of each species studied, their geographic distribution, and period of fruit maturation.

In 1994, a cooperative project between the Brasília Botanical Garden and Embrapa/Cenargen was established. An in vivo collection of medicinal plants from Cerrado, now contains 161 accessions. The collection has facilited phytochemical and pharmacological studies of this plant materials, and an anti-inflamatory agent has been identified on Lychnophora salicifolia.

Priority Species

A few germplasm collections of medicinal and aromatic plants have been established in

Brazil. The following species, listed alphabetically, have been recognized as priority for germplasm conservation.

Espinheira santa is a small shrub evergreen tree reaching up to 5 m height. It is native to many parts of southern Brazil, mainly in Paraná and Santa Catarina states.

Leaves of Maytenus species are used in the popular medicine of Brazil for their reported antiacid and antiulcerogenic activity. The effects of a boiling water extract of equal parts of Maytenus aquifolium and Maytenus ilicifolia leaves have been tested in rats and mice. Attempts to detect general depressant, hypnotic, anticonvulsant, and analgesic effects were reported by Oliveira et al. The potent antiulcerogenic effect of espinheira santa leaves was demonstrated effective compared to two leading anti-ulcer drugs, Ranitidine and Cimetidine. Toxicological studies demonstrated the plant's safety.

Seeds of Maytenus ilicifolia can be classified as orthodox and stored at -20 ℃ in long-term cold chambers. The Forestry Department of the University of Parana began a project in 1995 to study the genetic variability of natural populations of Maytenus ilicifolia and 78 accession were collected in the states of Parana, Santa Catarina, and Rio Grande do Sul. Field collections are maintained at the university campus. Although cultivation of Maytenus ilicifolia is the object of several studies in Brazil, a research focus on in situ conservation and sustainable systems of harvesting are required.

https://hort.purdue.edu/newcrop/proceedings1999/v4-152.html

Vocabulary

  biome 英 [ˈbaɪəʊm] 美 [ˈbaɪoʊm] *n.* (生态) 生物群系
  territory 英 [ˈterətri] 美 [ˈterətɔːri] *n.* 领地；领域；范围
  fragile 英 [ˈfrædʒaɪl] 美 [ˈfrædʒl] *adj.* 易碎的；脆的；虚弱的
  dominion 英 [dəˈmɪniən] 美 [dəˈmɪnjən] *n.* 统治权；领土；疆土；版图
  herbaceous 英 [hɜːˈbeɪʃəs] 美 [ɜːrˈbeɪʃəs] *adj.* 草本的；叶状的
  stratum 英 [ˈstrɑːtəm] 美 [ˈstreɪtəm] *n.* 地层；岩层；社会阶层
  arboreal 英 [ɑːˈbɔːriəl] 美 [ɑːrˈbɔːriəl] *adj.* 树木的；生活于树上的
  semi-arid climate 半干旱气候
  predominantly 英 [prɪˈdɒmɪnəntli] 美 [prɪˈdɑːmɪnəntli] *adv.* 占主导地位地
  mesophytic 英 [mesəʊˈfitik] 美 [mesəʊˈfitik] *adj.* 中植代的；中生植物的
  chamomile 英 [ˈkæməmaɪl] 美 [ˈkæməˌmaɪl, -ˌmil] *n.* 甘菊；黄春菊
  calendula 英 [kəˈlendjʊlə] 美 [kəˈlendʒələ] *n.* 金盏花属植物
  lemon balm 蜜蜂花
  rosemary 英 [ˈrəʊzməri] 美 [ˈroʊzmeri] *n.* [植]迷迭香
  oregano 英 [ˌɒrɪˈɡɑːnəʊ] 美 [əˈreɡənoʊ] *n.* 牛至
  sediment 英 [ˈsedɪmənt] 美 [ˈsɛdəmənt] *n.* 沉淀物
  genetic erosion 遗传冲刷；遗传侵蚀

ethnopharmacological　民族药理学
germplasm 英 ['dʒɜːmplæzm] 美 ['dʒɜːmplæzm] *n.* 种质；胚质
tissue culture　组织培养；体素培养
orthodox 英 ['ɔːθədɒks] 美 ['ɔːrθədɑːks] *adj.* 正统的；规范的
cryopreservation 英 [kraɪəprezə'veɪʃn] 美 [kraɪəprezə'veɪʃn] *n.* 冷冻保存；低温贮藏
sprout 英 [spraʊt] 美 [spraʊt] *n.* 幼芽；新梢；[植] 球芽甘蓝
herbarium voucher　植物标本记帐凭证
mutation 英 [mjuː'teɪʃn] 美 [mju'teʃən] *n.* 突变；变异；变化
vitro conservation　体外保存
anthropogenic 英 [ˌænθrəpə'dʒnɪk] 美 [ˌænθrəpə'dʒnɪk] *adj.* 人为的；人类活动产生的
foci 英 ['fəʊsaɪ] 美 ['foʊsaɪ] *n.* （病）灶
ethnobotanical 英 [ˌeθnəʊbə'tænəkəl] 美 [ˌeθnoʊbə'tænəkəl] *adj.* 民族植物学的
herbaria 英 [hɜː'beərɪə] 美 [hɜː'beərɪr] *n.* 腊叶集；干燥标本集（herbarium 的名词复数）
accession 英 [æk'seʃn] 美 [æk'sɛʃən] *n.* 增加；就职；增加物
ulcerogenic 英 [ˌʌlsərəʊ'dʒenɪk] 美 [ˌʌlsəroʊ'dʒenɪk] *adj.* 产生溃疡的

Useful Expressions

1. distribute 英 [dɪ'strɪbjuːt] 美 [dɪ'strɪbjut] *vt.* 分配；散布；散发；分发
Students shouted slogans and distributed leaflets.
学生们喊着口号，分发着传单。
2. initiate 英 [ɪ'nɪʃieɪt] 美 [ɪ'nɪʃieɪt] *vt.* 开始；发起
They wanted to initiate a discussion on economics.
他们想启动一次关于经济学的讨论。
3. as follows　列举如下
The reasons are as follows.
其理由如下。
4. adverse 英 ['ædvɜːs] 美 ['ædvɜːrs] *adj.* 不利的；有害的；逆的；相反的
The police said Mr. Hadfield's decision would have no adverse effect on the progress of the investigation.
警方称哈德菲尔德先生的决定将不会对调查的进展造成不利影响。

Questions

1. Why is Brazil considered the country with the greatest biodiversity?
2. What has been done to preserve the genetic variability of medicinal plants in Brazil?
3. Can you tell us how to store seeds of Maytenus ilicifolia?

Passage 2

Conservation Issues

If existing medicinal-plant resources are to continue to meet demand now and in the future, they will need to be adequately protected through the development of appropriate policies and legislation. Awareness of the conservation issues and of the importance of sustainable utilization needs to be raised among all stakeholders. Perhaps most importantly, local people need to be supported and encouraged to take the necessary steps to protect this valuable resource. The collection of medicinal plants must be guided by an accurate knowledge of the biology of the species concerned, and steps must be taken to avoid over-exploitation, and the collection of rare or otherwise endangered species.

Preserving Wild Genes. Fortunately, many plant species consist of thousands of populations. These together form a gene pool in which a more or less free gene exchange can take place. This is a feature that can be utilized by plant breeders to protect medicinalplant diversity.

As an example of the importance of preserving medicinal plants consider the case of silphion, a weed was once used as a contraceptive. It was apparently so effective that the Ancient Greeks literally revered it. Now, with population growth seemingly out of control a plant like this could have immense significance. Unfortunately, the Greeks used so much of it; it became extinct. Botanists can no longer find the species.

Between 570 and 250 BC the majority of coins minted in ancient Cyrene, a city situated in what is now the eastern part of Libya, carried the embossed picture of the Silphion plant. This reflects the enormous economic importance this plant had for the city over four centuries.

The perennial roots and strongly ribbed annual stems of the Silphion plant were eaten in the fresh state and were regarded as a perfume, flavoring agent and spice. The juice was employed medicinally against a wide range of symptoms and diseases, especially gynecological ailments—it was a true "multi-purpose species" in the sense of modem economic botany.

It appears that Silphion was found only in the dry hinterland. Attempts to cultivate it seem to have failed, so wild plants remained the source of supply. No reasons have been given for its disappearance although overharvesting is considered to be at least one reason for the dramatic decline in its use and final extinction as an economic resource. What we have is an example of overharvesting and probable extinction of an ancient medicinal plant. Silphion reflects both the potential wealth through plant utilization and the possible risks and downfall through overharvesting.

For historic (if not biological) reasons, the majority of medicinal plants used in developing countries are located in specific ecosystems. Prohibiting wild collections in these locations could devastate many poor families by cutting off their source of income. It is therefore important that education programs that justify the need for regulations governing in-situ conservation and collecting be developed. The local people should participate in this and the efforts should be linked to ex-situ conservation and cultivation programs that would provide an alternative source of income (or perhaps an equal income from smaller harvest through such means as improved quality control).

In-Situ Conservation. The protection of medicinal-plant resources was not identified as a major concern of conservation organizations until 1984. Four years later, the Chiang Mai Declaration recognized medicinal plants as an important component of the globe's biota. It noted that these plants are an essential part of primary healthcare in most of the world, and it viewed with alarm the rapidly increasing loss. The Global Biodiversity Strategy recognized the importance of conserving medicinal-plant biodiversity. Its socalled "Action 40" calls for the development of traditional medicines to ensure their appropriate and sustainable use, and "Action 41" promotes recognition of local knowledge, particularly medicinal healers. "Action 67" specifically mentions medicinal plants as a key group deserving increased attention. At the Rio Conference in 1992 the "Convention on Biological Diversity" ratified these action items.

Nonetheless, only a few countries seem to have pursued their obligations regarding medicinal-plant conservation. One of these is Sri Lanka, where the government has for a long time implanted in its people a strong pride in their natural heritage. Sri Lanka is a good example for other countries to follow. Its flora and fauna enjoy a high level of protection, with over 400 reserves set aside for their conservation. 14 Stringent laws apply in these reserves. The government has an aggressive policy of in-situ conservation to save valuable species, and in particular medicinal plants. This action was, in part, linked to the rapid resurgence of Ayurveda following independence and the demand for medicinal plants for Ayurvedic drugs. A Ministry of Indigenous Medicine was established in 1980. In 1986, the World Wide Fund for Nature (WWF) funded the Conservation of Medicinal Plants of Sri Lanka with the objective of establishing an aggressive policy of in situ conservation to save valuable species from extinction. The World Conservation Monitoring Center (WCMC) provides services to CITES. The CITES database is the largest of its kind, currently holding some two million entries on trade in wildlife species and their derivatives.

WCMC is the only organization that gathers, analyzes and provides information on plants threatened with extinction on a global scale. The centre is aware of the growing need to protect and conserve medicinal plants. Because of the potentially large number of medicinal plants requiring protection and the limited funds available categorizing medicinal-plant species, the following characteristics could be used to set priorities.

A partnership between WCMC and the World Bank established in 1995 will provide full biodiversity data mapping services to the World Bank: Seek to extend these services to GEF partners in UNEP and UNDP; Capture and mobilize data deriving from investments in biodiversity; Repatriate data to the developing world; build capacity for biodiversity information management in the developing world and strengthen information networks. Being able to access medicinal-plant data will enhance the decision-making process regarding protection, research priorities, management objectives, and polices to yield best results using ever scarce financial resources. It is on the basis of such information medicinal-plant diversity can be preserved in situ, successfully sustained, and ensure the germplasm for long-term ex-situ conservation and cultivation.

Ex-Situ Conservation. In 1989 the Botanic Gardens Conservation International (BGCI), in collaboration with the International Union for Conservation of Nature and WWF, published *The Botanic Gardens Conservation Strategy* as a guide for the development of botanic garden roles in biodiversity conservation. It has developed a computer database listing rare and endangered plants in cultivation in about 350 botanic gardens worldwide, which is used to foster networking and linkages. BGCI considers medicinal plants as a priority area for botanic gardens for the future. In July 1995, BGO launched an appeal for funds to establish an effective network of botanic gardens for medicinal plant ex-situ conservation and to strengthen the capacity of botanic gardens in developing countries. The first such gardens will be established in Colombia, Haiti, Uganda and Vietnam.

http: // www. nzdl. org/gsdlmod? e = d-00000-00---off-0cdl-00-0----0-10-0---0---0direct-10---4-------0-1l-11-en-50---20-preferences---00-0-1-00-0-0-11-1-0utfZz-8-00-0-0-11-10-0utfZz-8-10&a=d&c=cdl&cl=CL1. 86&d=HASH21fa548251018653c199d3. 6. 6

Vocabulary

stakeholder 英 ['steɪkhəʊldə(r)] 美 ['steɪkhoʊldə(r)] *n.* 利益相关者；股东
contraceptive 英 [ˌkɒntrə'septɪv] 美 [ˌkɑːntrə'septɪv] *n.* 避孕用具；避孕剂
revere 英 [rɪ'vɪə(r)] 美 [rɪ'vɪr] *vt.* 敬畏；崇敬；尊崇
mint 英 [mɪnt] 美 [mɪnt] *vt.* 铸造；发明或创造
emboss 英 [ɪm'bɒs] 美 [ɪm'bɔːs] *vt.* 装饰；浮雕（图案）
perennial 英 [pə'reniəl] 美 [pə'rɛniəl] *adj.* 多年生的；终年的；四季不断的
gynecological 英 [ˌgaɪnəkə'lɒdʒɪkəl] 美 [ˌgaɪnəkə'lɒdʒɪkəl] *adj.* 妇产科医学的
botany 英 ['bɒtəni] 美 ['bɑːtəni] *n.* 植物学
hinterland 英 ['hɪntəlænd] 美 ['hɪntərlænd] *n.* 腹地；内陆地区
devastate 英 ['devəsteɪt] 美 ['dɛvəˌstet] *vt.* 毁灭；破坏
in-situ 英 [ɪn'saɪtuː] 美 [ɪn'saɪtuː] *adj.* 原位的

ex-situ 异地
implant 英 [ɪmˈplɑːnt] 美 [ɪmˈplænt] *vt.* 移植
fauna 英 [ˈfɔːnə] 美 [ˈfɔːnə] *n.* 动物群；动物志
propagation 英 [ˌprɒpəˈgeɪʃn] 美 [ˌprɑpəˈgeʃən] *n.* 宣传；[生]繁殖法
botanic 英 [bəˈtænɪk] 美 [bəˈtænɪk] *adj.* 植物的；植物学的

Useful Expressions

1. preserve 英 [prɪˈzɜːv] 美 [prɪˈzɜːrv] *vt.* 保护；保持；保存
We will do everything to preserve peace.
我们会尽全力维持和平。
2. rib 英 [rɪb] 美 [rɪb] *vt.* 嘲笑；开（某人的）玩笑
The guys in my local pub used to rib me about drinking "girly" drinks.
我家当地酒馆里的人以前总是取笑我喝"女人"饮料。
3. mobilize 英 [ˈməʊbəlaɪz] 美 [ˈmoʊbəlaɪz] *vt.* 调动；动员
The best hope is that we will mobilize international support and get down to action.
最好的希望是我们能争取到国际上的支持，然后采取行动。
4. repatriate 英 [ˌriːˈpætrieɪt] 美 [ˌriːˈpeɪtrieɪt] *vt.* 回国；遣返
It was not the policy of the government to repatriate genuine refugees.
遣返真正的难民回国并非政府的政策。
5. strengthen 英 [ˈstreŋθn] 美 [ˈstrɛŋkθən, ˈstrɛŋ-, ˈstrɛn-] *vt.* 加强；巩固
To strengthen his position in Parliament, he held talks with leaders of the Peasant Party.
为巩固其在国会中的地位，他曾与农民党的领导人进行多次会谈。

Questions

1. Why is it necessary to protect medicinal plants? What must we do?
2. What is the use of silphion?
3. What can we get from the partnership between WCMC and the World Bank?

3.2 Reserves of Main Medicinal Plants Resources

Passage 1

Notable Chinese Medicinal Plants

During the past 30 years, the identifications of the historically recorded medicinal plants have been verified and their chemical taxonomy determined. *The Encyclopedia of Traditional Chinese Crude Drugs* describes the botanical and analytical standards of 5646 crude drugs. The latest edition of the *Chinese Pharmacopoeia* contains a list of 647 crude drugs of botanical origin, their formulations, methods of preparation, requirements and tests for strength and purity, and related information. The Ministry of Health has begun the standardization of the names of all phytopharmaceutical preparations.

Three of the most commonly-used plant species in Chinese medicinal preparations are described below. They give a sense of the botanical wealth to be found in China's natural resource heritage.

Ginseng (Panax ginseng). Probably the most famous among Chinese traditional drugs, ginseng was first described in *Compendium of materia medica* written almost 2,000 years ago in china. It is by the 4th century, centers of production, time of harvesting and morphological characters had been recorded. During the past 1,500 years, the value of ginseng has remained high—"equal to its weight in silver". The plant also occurs and is cultivated in Korea, Japan, Russia and North America. Because the root shape can resemble the human form, it was believed to be effective in curing disease and strengthening the weak (i. e. a general cardiac tonic). Its medicinal value appears to stimulate the pituitary gland resulting in homeostasis (chemical and metabolic balance). This concept is the central principle of traditional Chinese medicine.

The price while high in 1988 ($250,000 per ton) dropped in 1989 with record yields to $80,000 per ton, but has been rebounding in recent years (for instance, in 1995 it was $150,000 per ton). The significant price difference between 1988 and 1989 might be indicative of what can happen with increased production and no price control at the farm level. Approximately 2,000 tons of dried ginseng with a value of $50,000 per ton are exported annually. Another 2,000 tons (undocumented) are also exported. At the same time North American ginseng exports to China doubled between 1993 (1,140 tons) and 1995 (2,200 tons).

Eucommia (Eucommia ulmoides). This plant also known as the gutta-percha tree, has been an important economic plant and is endemic to the mountainous regions of China. It is now known only in cultivation, having been harvested into extinction in the wild. All parts of the tree are valuable but the bark is the main medicinal. For many centuries, eucommia, bark or tu-chung was used traditionally as a rejuvenating tonic to benefit the liver and kidney, and to strengthen the muscles and bones. It was only in 1948 that its antihypertensive activity was discovered. The bark is the source of the active compound pinoresinal di-D-glucoside.

This tree's bark, fruit, and leaves contain 6 to 18 percent gutta-percha, a material chemically akin to natural rubber but that is hard and lacks "bounce". The extracted rubber has excellent insulation properties, low moisture absorption and is resistant to acid, alkali, oil, and corrosion, and represents one of the important raw materials for the manufacturing of undersea cables and airplane tires. It has excellent bonding properties, serving as materials for filling teeth and setting bones. The seed is the source of high quality cooking oil. The leaves contain vitamin C and may be used as tea. The wood is valued for manufacturing furniture and handicrafts.

The tree is found in more than 260 counties of 16 Chinese provinces. Hunan is the major center of production, producing more than all the other provinces together. The Province has the Eucommia Scientific Research Centre located in Cili County. Approximately 0.2 million hectares are under eucommia plantations. The total annual yield is about 4,000 tons of bark, of which about 2,000 tons are exported. Production is expected to reach 5,000 tons by 2,000. While leaf production is more difficult to calculate, exports in 1993 reached 5,000 tons.

Despite the quantities produced, there is not enough to meet the demand. Because of its many uses the bark's market prices are high and stable; domestic prices are between $6 per kilogram and the international market is $80 per kilogram. 35 production is expected to increase significantly in the future as the plant can be intercropped with food crops and used to rehabilitate degraded hillsides.

Seabuckthorn (Hippophae rhamnoides). Human beings have been using this shrub for at least 1,200 years. The plant known in English as seabuckthorn, was recorded in the Tibetan medicinal classics (*the Four Books of Pharmacopoeia*) completed in the Tang Dynasty (618—907 AD). Although China was one of the earliest countries in the world to use seabuckthorn as a medicinal plant, until 1980 its use was limited to Tibet and Mongolia. The processing of seabuckthorn medicinal products did not start in China until 1986. It has proven to be a profitable crop because of its many uses in the medicinal, food, and cosmetic industries.

At present, 1.2 million hectares (95 percent of world total) of seabuckthorn are under cultivation in 19 provinces. Seven breeding stations have been established to select new varieties adapted to different biogeographic regions.

In China, there are an estimated 740,000 hectares and 300,000 hectares of natural and cultivated plants. As of 1995, more than 10,000 people were employed on various aspects of plant development, 95 percent are located in rural areas and do not include farmers. Because major economic benefits can be realized quickly (in three or four years) farmers are keen to plant. Approximately 50,000 tons of seabuckthorn berries are harvested annually and processed into 200,000 tons of various products valued at $35.7 million. The Chinese government has invested more than $25 million in seabuckthorn research and development.

The shrub has attracted a great deal of attention from scientists and engineers around the world because of its combined ecological and economic benefits. The seabuckthorn root system, for example, is so extensive that its roots can branch many times in a growing season and form a complex underground network that holds the soil from slippage like wire reinforcing mesh in concrete. When plants are buried under sediments massive adventitious roots extend to form new horizontal root systems. An individual plant can propagate massive bushes or a small forest in several years. This is why the seabuckthorn bushes play such a prominent role in protecting river banks, preventing floods and minimizing slope erosion. The plants are considered more effective than any construction work. Furthermore, its role in rehabilitation and upgrading of marginal or fragile slopes through soil-binding is well documented.

Where land degradation and its accompanying poverty occur it can play an important role in soil and water conservation and land rehabilitation. Seabuckthorn is a multipurpose plant, and its potential is far from fully exploited. With further study, more and more uses could be developed in the near future. Its humanitarian and economic benefits can be summarized as follows. The plant is

• a source of low-priced vitamins, seabuckthorn fruits can benefit millions of children suffering from vitamin A deficiency.

• a means for generating cash income, it has since 1985, in the middle reaches of the Yellow River, provided farmers with earnings of about $1.06 million from the sale of fruit every year.

• an option for stabilizing mountain slopes it is selected by farmers and engineers because of its extensive root system, soil binding qualities, its provision of good surface cover, and its utility as fodder, food, fuelwood, and supplier of medicine.

It seems no wonder, therefore, that a 1990 assessment put China's total area of seabuckthorn at about 1 million hectares, and the total value of its products at more than $20 million per year. Moreover, between 1991—1995, an additional 330,000 hectares were scheduled to be bought under seabuckthorn cultivation.

https://www.baidu.com/#ie=utf-8&f=3&rsv_bp=1&rsv_idx=1&tn=baidu &wd=Reserves%20of%20Main%20Medicinal%20Plants%20Resources&oq=Reserves% 20of%20Main%20Medicinal%20Plants%20Resources&rsv_pq=8a1f91c7000216f5&rs v_t=ebb93UyUJshQrk4aofRA68aKii6H689dTnXXHfy0xW5yDilDoLoYGfiFC3c&rqla ng=cn&rsv_enter=0&rsv_sug3=36&rsv_sug1=20&rsv_sug7=100&prefixsug=Res erves%2520of%2520Main%2520Medicinal%2520Plants%2520Resources&rsp=3&rsv_s ug4=6069&rsv_sug=1&pn=30&usm=1&rsv_page=1&bs=Reserves%20of%20Ma in%20Medicinal%20Plants%20Resources

Vocabulary

analytical 英 [ænəˈlɪtɪkl] 美 [ænəˈlɪtɪkl] adj. 分析的;分析法的
Chinese Pharmacopoeia 《中国药典》
standardization 英 [ˌstændədaɪˈzeɪʃn] 美 [ˌstændədaɪˈzeɪʃn] n. 规范化
phytopharmaceutical 植物药理学
cardiac 英 [ˈkɑːdiæk] 美 [ˈkɑːrdiæk] adj. 心脏(病)的;(胃的)贲门的

pituitary gland 英 [pɪˈtuːɪˌteri: glænd] 美 [pɪˈtuɪˌtɛri glænd] *n*. 脑下垂体
homeostasis 英 [ˌhəʊmiəˈsteɪsɪs] 美 [ˌhomioˈstesɪs] *n*. 原状稳定
rebound 英 [rɪˈbaʊnd] 美 [ˈriˈbaʊnd, rɪ-] *vt*. 弹回
North American ginseng　北美人参；西洋参
eucommia 英 [ˈjuːkəmɪə]　美 [ˈjuːkəmɪr] *n*. 杜仲
gutta-percha 英 [ˌgʌtə ˈpɜːtʃə] 美 [ˈpɜːrtʃə] *n*. 古塔胶；杜仲胶
rejuvenating tonic　滋补补药
antihypertensive 英 [ˈæntiːhaɪpəˈtensɪv] 美 [ˈæntiːhaɪpəˈtensɪv] *adj*. 抗高血压的（药物）
insulation 英 [ˌɪnsjuˈleɪʃn] 美 [ˌɪnsəˈleɪʃn] *n*. 绝缘；隔声；隔离；孤立
corrosion 英 [kəˈrəʊʒn] 美 [kəˈroʒən] *n*. 腐蚀；侵蚀；锈蚀
intercrop 英 [ˌɪntəˈkrɒp] 美 [ˌɪntəˈkrɒp] *v*. 间作
rehabilitate 英 [ˌriːəˈbɪlɪteɪt] 美 [ˌrihəˈbɪlɪˌtet] *vt*. 使康复；使复原；修复
degraded hillside　退化山坡
cosmetic 英 [kɒzˈmetɪk] 美 [kɑːzˈmetɪk] *n*. 化妆品；装饰品
hectare 英 [ˈhekteə(r)] 美 [ˈhekter] *n*. 公顷
slippage 英 [ˈslɪpɪdʒ] 美 [ˈslɪpɪdʒ] *n*. 滑动；下跌；滑动量；下跌量
mesh 英 [meʃ] 美 [mɛʃ] *n*. 网状物；陷阱
mesophytic 英 [ˌmesəʊˈfɪtɪk] 美 [ˌmesəʊˈfɪtɪk] *adj*. 中植代的；中生植物的

Useful Expressions

1. crude 英 [kruːd] 美 [krud]　*adj*. 简陋的；粗糙的；天然的；未加工的
Standard measurements of blood pressure are an important but crude way of assessing the risk of heart disease or strokes.
标准的血压测量是评估患心脏病或中风概率的一个重要但不精确的方法。
2. deficiency 英 [dɪˈfɪʃnsi] 美 [dɪˈfɪʃənsi] *n*. 缺乏；不足；缺点；缺陷；不足额
It's a chicken and egg situation. Does the deficiency lead to the eczema or has the eczema led to certain deficiencies?
这是个鸡和蛋的问题。是因为营养缺乏导致湿疹呢，还是湿疹导致了某些营养缺乏？
3. reinforce 英 [ˌriːɪnˈfɔːs] 美 [ˌriːɪnˈfɔːrs] *vt*. 加固；强化；增援
A stronger European Parliament would, they fear, only reinforce the power of the larger countries.
他们担心一个更加强大的欧洲议会只会增强大国的力量。

Questions

1. What are the most commonly-used plant species in Chinese medicinal preparations?
2. Which is the most famous among Chinese traditional drugs?
3. What part of eucommia is used as the main medicinal?
4. Why do scientists and engineers pay so much attention to seabuckthorn?

Passage 2

Predicting the Global Potential Distribution of Four Endangered Panax Species in Middle- and Low-Latitude Regions of China by the Geographic Information System for Global Medicinal Plants (Extract)

Introduction

Panax (Araliaceae) species are the important medicinal resources in the world. Panax ginseng and Panax quinquefolium are distributed at high-latitude regions. By contrast, some endangered Panax species such as Panax japonicus and Panax japonicus var. major found in middle and low-latitude regions, have been recorded in the Chinese Pharmacopoeia. Panax japonicus is known as "the king of herbs" in Chinese folk, and has been included in the *Japanese Pharmacopoeia* as a traditional Japanese medicine. Panax zingiberensis and Panax stipuleanatus distributed in low-latitude regions are also used widely as traditional ethnic medicines. Panax species are popular due to their potential medicinal properties, such as anti-fatigue, anti-tumor, anti-thrombotic, anti-inflammatory, anti-oxidative, and immune-enhancing effects and thus have a substantial market demand worldwide. However, their resources are gradually declining due to excessive harvesting and lack of environmental protection. Therefore, predicting the distribution of plants is necessary for their conservation.

With the development of network technology, a distribution prediction model has become one of the common methods in biodiversity conservation, such as MaxEnt (Maximum Entropy), Random Forset, GMPGIS (the geographic information system for global medicinal plants). GMPGIS selected seven key ecological factors (especially related to its growth and accumulation of secondary metabolism) for medicinal plants according to biological characteristics of medicinal plants, statistics, ecology, botany, and related literatures, and experience. In this research, the global potential distribution of four endangered Panax species found in middle-and low-latitude regions was predicted by using GMPGIS. GMPGIS model has been created by the Institute of Chinese materia medica in order to predict the distribution of medicinal plants using environment databases such as WorldClim, CliMond, and HWSD. This model has been used for introduction and conservation of Panax ginseng C. A. Mey and Panax notoginseng (Burk.) F. H. Chen. By combining the climate and soil factors to explore areas that have the most similar ecological factors, we can determine the suitable environment for medicinal plants to scientifically protect and cultivate the endangered plants. MaxEnt is also widely used to predict species distribution. Potential suitable distribution plays an important role in resource protection and cultivation of endangered plants. We studied and compared the differences between GMPGIS and MaxEnt and found that GMPGIS shows higher precision than MaxEnt. In this research, GMPGIS was used to analyze the plant habitats, whereas MaxEnt was used for supplementary information.

We used GMPGIS to scientifically predict the global distribution of four endangered Panax species based on the climate and soil factors. The global potential suitable distributions were mapped by excluding the unsuitable areas, such as lakes, rivers, and cities. This research aims to predict the global potential suitable distribution of four endangered Panax species and provide a scientific reference for protecting wild resources and breeding endangered plants.

Results

1) Ecological Factors

GMPGIS was used to extract the ecological factor data of collected sampling points. Panax zingiberensis and Panax stipuleanatus are distributed in low-latitude regions which include China and Burma. Panax japonicus and Panax japonicus var. major are distributed in middle-and low-latitude regions. According to some figures, it shows the whole ecological factors range of each plant which plays an important role in cultivation. The contributions of each ecological factor were revealed by MaxEnt. We found that the proportions of the mean temperature of the coldest quarter were respectively 44.7%, 55.5%, 45.6%, and 43.0% for Panax japonicus var. major, Panax japonicas, Panax zingiberensis, Panax stipuleanatus. The proportions of annual precipitation were 36.0%, 37.8%, and 42.3% for P. japonicas, Panax zingiberensis, Panax stipuleanatus. This result provides a favorable condition for further analysis and experiment, and we can clearly understand the suitable range of climate factors and soil for these endangered medicinal plants.

2) Potential Distribution

(1) Global Potential Distribution

The potential distribution of endangered plants was influenced by the variation in the soil and climatic factors. Some environmental variables were also strongly correlated with the potential distributions. a. Potential distribution of Panax japonicus: The global potential distribution of Panax japonicus is obtained by GMPGIS based on the ecological factors and soil, and the total area is 118.29×10^5 km^2. Panax japonicus is distributed mainly in North and South America, Asia, Europe, Oceania, and other regions. The leading distribution areas are Southeast Asia and North America, which include China, Japan, Korea, the United States, and Canada. The top three distribution areas are China ($2,662.98 \times 10^3$ km^2), the United States ($2,312.34 \times 10^3$ km^2), and France (260.81×10^3 km^2). b. Potential distribution of Panax japonicus var. major: The potential distribution regions for this plant are found in North America, Asia, and Europe. The total global potential distribution area is 77.5×10^5 km^2, and the top three distribution areas are the United States ($3,438.73 \times 10^3$ km^2), China ($2,986.11 \times 10^3$ km^2), and

Russia (861.09×10^3 km^2). c. Potential distribution of Panax zingiberensis: The global potential suitable distribution of Panax zingiberensis is obtained by GMPGIS. This species is distributed in a small part of Asia and South America. The total global potential suitable distribution area is 5.09×10^5 km^2, the top three distribution areas are Brazil (232.79×10^3 km^2), China (166.71×10^3 km^2), and the United States (39.58×10^3 km^2). Thus, the introduction and cultivation of Panax zingiberensis should be prioritized in these countries and regions. d. Potential distribution of Panax stipuleanatus: The global potential suitable distribution of this plant is limited to several countries in Asia and South America, and the total area is 2.05×10^5 km^2. The top three distribution areas are China (108.03×10^3 km^2), Brazil (35.92×10^3 km^2), and Burma (27.33×10^3 km^2).

(2) Chinese Potential Distribution

Panax japonicus and Panax japonicus var. major are mainly distributed in the middle-latitude regions of China, including Sichuan Province, Guizhou Province, Shanxi Province, Shandong Province, Hebei Province, and Yunnan Province. However, the ecological adaptation areas of Panax zingiberensis and Panax stipuleanatus are limited and mainly distributed in Yunnan Province, Guangdong Province, and Fujian Province. In addition, Panax japonicus C. A. Mey is the most widely distributed species with ecological adaptation area of $2,986.11 \times 10^3$ km^2 in China.

http://www.mdpi.com/1420-3049/22/10/1630/html

Vocabulary

Panax 英 ['peinæks] 美 ['peinæks] 人参
Panax ginseng 人参
accumulation of secondary metabolism 次生代谢积累

Useful Expressions

1. correlate 英 ['kɔrileit] 美 ['kɔrə,let] v. (使)相同于；把……联系起来
His research results correlate with yours.
他的研究成果和你的研究成果有关联。
2. prioritize 英 [praɪ'ɒrətaɪz] 美 [praɪ'ɔːrətaɪz] vt. 优先处理
Make lists of what to do and prioritize your tasks.
将要做的事情列出来，然后决定其优先顺序。

Questions

1. How many Panax species are mentioned in the passage?
2. How do ecological factors influence Panax?
3. What influences the distribution of Panax?

Chapter Four
Resources Chemistry of Medicinal Plants

4.1 Overview

Passage 1

Chemical Markers for the Quality Control of Herbal Medicines: An Overview (Part 1)

Background

Herbal medicines, also known as botanical medicines or phytomedicines, refer to the medicinal products of plant roots, leaves, barks, seeds, berries or flowers that can be used to promote health and treat diseases. Medicinal use of plants has a long history worldwide. According to the World Health Organization (WHO), traditional herbal preparations account for 30%—50% of the total medicinal consumption in China. There have always been concerns about the inconsistent composition of herbal medicines and occasional cases of intoxication by adulterants and/or toxic components. Quality control of herbal medicines aims to ensure their consistency, safety and efficacy.

Chemical fingerprinting has been demonstrated to be a powerful technique for the quality control of herbal medicines. A chemical fingerprint is a unique pattern that indicates the presence of multiple chemical markers within a sample.

The European Medicines Agency (EMEA) defines chemical markers as chemically defined constituents or groups of constituents of a herbal medicinal product which are of interest for quality control purposes regardless whether they possess any therapeutic activity. Ideally, chemical markers should be unique components that contribute to the therapeutic effects of a herbal medicine. As only a small number of chemical compounds were shown to have clear pharmacological actions, other chemical components are also used as markers. The quantity of a chemical marker can be an indicator of the quality of a herbal medicine.

The overall quality of a herbal medicine may be affected by many factors, including seasonal changes, harvesting time, cultivation sites, post-harvesting processing, adulterants or substitutes of raw materials, and procedures in extraction and preparation. From harvesting to manufacturing, chemical markers play a crucial role in evaluating the quality of herbal medicines. Moreover, the study of chemical markers is applicable to many research areas, including authentication of genuine species, search for new resources or substitutes of raw materials, optimization of extraction and purification methods, structure elucidation and purity determination. Systematic investigations using chemical markers may lead to discoveries and development of new drugs.

In this review, we summarise selection criteria for chemical markers and how chemical markers are used to evaluate the quality of herbal medicines.

Selection of Chemical Markers

A total of 282 chemical markers are listed in the *Chinese Pharmacopoeia* (2005 edition) for the quality control of Chinese herbal medicines. As discussed in the monographs of the *American Herbal Pharmacopoeia*, the use of single or multiple chemical markers was important to quality control. Scientists and regulatory agencies have paid attention to the selection of chemical markers in quality control. The EMEA categorises chemical markers into analytical markers and active markers. According to the definition by the EMEA, analytical markers are the constituents or groups of constituents that serve solely for analytical purposes, whereas active markers are the constituents or groups of constituents that contribute to therapeutic activities. There are other classifications of chemical markers. For example, Srinivasan proposed the following four categories: active principles, active markers, analytical markers and negative markers. Active principles possess known clinical activities; active markers contribute to clinical efficacy; analytical markers have no clinical or pharmacological activities; negative markers demonstrate allergenic or toxic properties. All markers may contribute to the evaluation, standardisation and safety assessment of herbal medicines. Lin et al. expanded Srinivasan's classification into seven categories, namely, active principles, active markers, group markers, chemical fingerprints, analytical markers, "phantom" markers and negative markers. Group chemical markers have similar chemical structures and/or physical properties. The pharmacological activities of individual components are not necessarily known. Polysaccharides are classified under this category. This type of markers is not necessarily specific and can be easily masked by other components especially in proprietary products. Phantom markers are constituents that have known pharmacological activities; however, they can be undetectable in some herbal medicines due to low quantities. Special care should be taken when "phantom" markers were selected as chemical markers for quality control. While group chemical markers have a lower resolving power in qualitative analysis, chemical fingerprinting cannot provide adequate quantitative information.

Here, we suggest a new classification of eight categories of chemical markers, namely, (i) therapeutic components, (ii) bioactive components, (iii) synergistic components, (iv) characteristic components, (v) main components, (vi) correlative components, (vii) toxic components, and (viii) general components used with fingerprint spectrum. These eight categories are defined and discussed in the subsequent sections.

Therapeutic Components

Therapeutic components possess direct therapeutic effects of a herbal medicine. They may be used as chemical markers for both qualitative and quantitative assessments.

Originated from the bulbs of Fritillaria species (family Liliaceae), Bulbus Fritillariae (Beimu) is commonly prescribed as an antitussive and expectorant in Chinese medicine practice. Five different Bulbus Fritillariae derived from nine Fritillaria species are documented in the *Chinese Pharmacopoeia*. Isosteroidal alkaloids of Bulbus Fritillariae, including verticine, verticinone and imperialine, were identified as the major therapeutic components that account for the antitussive effect. Therefore, isosteroidal alkaloids were selected as the chemical markers for the quality assessment of Bulbus Fritillariae using a series of chromatographic techniques such as pre-column derivatizing gas chromatography-flame ionization detection (GC-FID), direct GC-FID, gas chromatography-mass spectrometry (GC-MS), pre-column derivatizing high-performance liquid chromatography-ultraviolet detection (HPLC-UV), high-performance liquid chromatography-evaporative light scattering detection (HPLC-ELSD) and high-performance liquid chromatography-mass spectrometry (HPLC-MS) methods.

Artemisinin from Herba Artemisiae Annuae (Qinghao) is another example of therapeutic component. Herba Artemisiae Annuae is well known for its potent anti-malarial activity. Artemisinin inhibits Plasmodium falciparum and Plasmodium vivax, two pathogens that cause malaria. Artemisinin is now used as a chemical marker in HPLC-ELSD, GC-FID and GC-MS for assessing the quality of the plant (parts and whole) at various stages, including the green and dead leaves of the plant.

Bioactive Components

Bioactive components are structurally different chemicals within a herbal medicine; while individual components may not have direct therapeutic effects, and the combination of their bioactivities does contribute to the therapeutic effects. Bioactive components may be used as chemical markers for qualitative and quantitative assessment.

According to Chinese medicine theories, Radix Astragali (Huangqi), derived from the roots of Astragalus membranaceus (Fish.) Bge. or A. membranaceus var. mongholicus (Bge.) Hsiao, is used to reinforce qi. Isoflavonoids, saponins and polysaccharides of Radix Astragali showed pharmacological actions in immune and circulatory systems, which were consistent with the Chinese medicine indications. These bioactive components, including isoflavonoids and saponins, were used simultaneously in the evaluation of the quality of Radix Astragali.

Synergistic Components

Synergistic components do not contribute to the therapeutic effects or related bioactivities directly. However, they act synergistically to reinforce the bioactivities of other components, thereby modulating the therapeutic effects of the herbal medicine. Synergistic components may be used as chemical markers for qualitative and quantitative assessment.

The products of St. John's wort (Hypericum perforatum L.) are popular for treating mild depression. Butterweck et al. reviewed the research progress on the phytochemistry and pharmacology of St. John's wort. Naphthodianthrone, hypericin, and hyperforin (a phloroglucinol derivative) were identified as the major components that contribute to the pharmacological activities of St. John's wort. Rutin, a ubiquitous flavonoid of natural products, demonstrated synergistic antidepressant actions in St. John's wort. In a forced swimming test on rats, extracts of St. John's wort with various chemical profiles were tested, among which the extract containing about 3% of rutin showed positive effects, whereas the extracts containing less than 3% of rutin were inactive. The extracts became active when the level of rutin was increased to about 3%. However, rutin alone did not show any effects under the same conditions. These results suggest that chemicals in St. John's wort work synergistically to achieve the antidepressant effects. Therefore, naphthodianthrones, phloroglucinols and flavonoids may be used as chemical markers for the quality control of St. John's wort.

Characteristic Components

While characteristic components may contribute to the therapeutic effects, they must be specific and/or unique ingredients of a herbal medicine.

Terpene lactones in the leaves of Ginkgo biloba L. (Yinxing) exemplify characteristic components. EGb 761, a standardized leave extract of Ginkgo biloba is a well defined product for the treatment of cardiovascular diseases, memory loss and cognitive disorders associated with age-related dementia. Flavonoids and terpene lactones are responsible for the medicinal effects of EGb 761. Flavonoids, terpene lactones including ginkgolides A, B and C, and bilobalide are chemical markers for the quality control of Ginkgo biloba leave extracts. EGb 761 contains 6% of terpene lactones (2.8%—3.4% of ginkgolides A, B and C, and 2.6%—3.2% of bilobalide) and 24% of flavone glycosides. Aglycons are primarily quercetin, kaempferol and isorhamnetin.

Valerenic acids, the characteristic components of valerian derived from the roots of Valeriana officinalis L., have sedative effects and improve sleep quality. Valerenic acids are used as chemical markers to evaluate the quality of valerian preparations although their sedative effects have not been fully elucidated. These chemical markers are also used for studying in vitro release of coated and uncoated tablets and stability test for valerian ground materials and extracts.

https://cmjournal.biomedcentral.com/articles/10.1186/1749-8546-3-7

Vocabulary

phytomedicine   *n.* 植物药
traditional herbal preparations   中药制剂
inconsistent 英 [ˌɪnkən'sɪstənt] 美 [ˌɪnkən'sɪstənt] *adj.* 不一致的
intoxication 英 [ɪnˌtɒksɪ'keɪʃn] 美 [ɪnˌtɑksɪ'keʃən] *n.* 陶醉
adulterant 英 [ə'dʌltərənt] 美 [ə'dʌltərənt] *n.* 掺杂物
efficacy 英 ['efɪkəsi] 美 ['ɛfɪkəsi] *n.* 功效；效力
chemical fingerprinting   指纹图谱
constituent 英 [kən'stɪtjuənt] 美 [kən'stɪtʃuənt] *n.* 选民；成分；构成部分
authentication 英 [ɔːˌθentɪ'keɪʃn] 美 [ɔːˌθentɪ'keɪʃn] *n.* 认证；身份验证；鉴定
optimization 英 [ˌɒptɪmaɪ'zeɪʃən] 美 [ˌɑptəmɪ'zeʃən] *n.* 最佳化；最优化
extraction 英 [ɪk'strækʃn] 美 [ɪk'strækʃən] *n.* [化] 提取(法)；萃取(法)；提出物
allergenic 英 [əlɜː'dʒenɪk] 美 [əlɜ'dʒenɪk] *adj.* 引起变态反应的
proprietary 英 [prə'praɪətri] 美 [prə'praɪəteri] *n.* 专卖药品；独家制造(及销售)的产品
category 英 ['kætəɡəri] 美 ['kætəɡɔːri] *n.* 类型；种类
bioactive component   生物活性成分
spectrum 英 ['spektrəm] 美 ['spɛktrəm] *n.* 光谱；波谱；范围；系列
antidepressant 英 [ˌæntɪdɪ'presnt] 美 [ˌæntɪdɪ'presənt, ˌæntaɪ-] *n.* 抗抑郁剂
valerian 英 [və'lɪəriən] 美 [və'lɪriən] *n.* 缬草

Useful Expressions

1. account for   说明(原因、理由等)；导致；引起
Now, the gene they discovered today doesn't account for all those cases.
他们现在发现的基因无法解释所有的病例。
2. consumption 英 [kən'sʌmpʃn] 美 [kən'sʌmpʃən] *n.* 消费；肺病；耗尽
The laws have led to a reduction in fuel consumption in the US.
这些法律已经使美国燃料消费量有所减少。
3. ensure 英 [ɪn'ʃʊə(r)] 美 [ɛn'ʃʊr] *vt.* 确保；使(某人)获得；使安全
Britain's negotiators had ensured that the treaty which resulted was a significant change in direction.
英国谈判代表已经担保由此而签订的条约将是方向上的重大改变。

Questions

1. What is phytomedicine?
2. What is chemical marker according to EMEA?
3. Why are the eight chemical markers selected?

Passage 2

Chemical Markers for the Quality Control of Herbal Medicines: An Overview (Part 2)

Main Components

Main components are the most abundant in a herbal medicine (or significantly more abundant than other components). They are not characteristic components and their bioactivities may not be known. Main components may be used for both qualitative and quantitative analysis of herbal medicines especially for differentiation and stability evaluation.

Four well-known Chinese herbal medicines derived from the genus Panax, namely (i) Radix et Rhizoma Ginseng (Renshen), (ii) Radix et Rhizoma Ginseng Rubra (Hongshen), (iii) Radix Panacis Quinquefolii (Xiyangshen) and (iv) Radix et Rhizoma Notoginseng (Sanqi), contain triterpenoid saponins including ginsenoside Rg1, Re, Rb1 and notoginsenoside R1 as their main components. Through qualitative and quantitative comparison of the saponin profiles, these four herbs can be differentiated from one another.

Herba Epimedii (Yinyanghuo), derived from the aerial parts of Epimedium brevicornum Maxim., E. sagittatum (Sieb. et Zucc.) Maxim. E. pubescens Maxim., E. wushanense T. S. Ying or E. koreanum Nakai, has been traditionally used to reinforce kidney-yang (Shenyang), strengthen tendons and bones, and relieve rheumatic conditions. Flavonoids including epimedin A, B, C and icariin are the main components of Herba Epimedii. A 24-month randomised, double-blinded and placebo-controlled clinical study showed that flavonoids from Herba Epimedii exerted beneficial effects on preventing osteopenia in late post-menopausal women. According to the *Chinese Pharmacopoeia* (2005 edition), total flavonoids and icariin are used as chemical markers for Herba Epimedii. In a recent study, Chen et al. developed an HPLC method to simultaneously quantify up to 15 flavonoids, of which epimedin A, B, C and icariin were selected as chemical markers for the quality assessment of the Epimedium species documented in the *Chinese Pharmacopoeia* (2005 edition).

Correlative Components

Correlative components in herbal medicines have close relationship with one another. For example, these components may be the precursors, products or metabolites of a chemical or enzymatic reaction. Correlative components can be used as chemical markers to evaluate the quality of herbal medicines originated from different geographical regions and stored for different periods of time.

According to the *Chinese Pharmacopoeia* (2005 edition), only psoralen and isopsoralen are used as chemical markers for assessing the quality of Fructus Psoraleae (Buguzhi). Recently, our group identified glycosides, psoralenoside and isopsoralenoside, and found them useful as the chemical markers for Fructus Psoraleae. The levels of psoralen and isopsoralen are inversely correlated to the level of glycosides psoralenoside and isopsoralenoside. When extracted with 50% methanol, samples had high levels of psoralen and isopsoralen and a minute amount of psoralenoside and isopsoralenoside. When moistened with 100% methanol, dried and extracted with 50% methanol, samples contained all four components. Psoralen and isopsoralen may be the enzymatic reaction products of psoralenoside and isopsoralenoside respectively. After incubation of a psoralenoside and isopsoralenoside solution with β-glucosidase at 36℃ for 24 hours, the amount of psoralen and isopsoralen became detected. Psoralenoside and isopsoralenoside, therefore, may be used as chemical markers for the quality control of Fructus Psoraleae.

Toxic Components

traditional Chinese medicine literature and modern toxicological studies documented some toxic components of medicinal herbs. For instance, aristolochic acids (AAs) and pyrrolizidine alkaloids (PAs) may cause nephrotoxicity and heptotoxicity respectively.

The use of three herbal medicines that contain AAs, namely Radix Aristolochiae Fangchi (Guangfangji), Caulis Aristolochiae Manshuriensis (Guanmutong) and Radix Aristolochiae (Qingmuxiang), have been prohibited in China since 2004. These three herbs were traditionally used to relieve pain and treat arthritis. Radix et Rhizoma Asari (Xixin) was traditionally sourced from the whole plants of Asarum heterotropoides Fr. Schmidt var. mandshurcum (maxim.) Kitag., A. sieboldii Miq. var. seoulense Nakai or A. seiboldii Miq. Its medicinal use has now been officially limited to the roots and rhizomes because the roots and rhizomes contain a much lower level of AAs than the aerial parts. AAs are now used as markers to control nephrotoxic herbs and proprietary herbal products.

There are over 6,000 plant species containing PAs which can cause hepatic veno-occlusive disease. There have been various PA restriction guidelines issued by government bodies and organisations. According to the WHO guidelines issued in 1989, the lowest intake rate of toxic PAs that may cause veno-occlusive disease in human is 15 μg/kg/day. In 1993, the American Herbal Products Association (AHPA) alerted its members to restrict the use of comfrey, a herbal medicine that contains PAs for external applications. In 2001, the Food and Drug Administration (FDA) of the United States recalled comfrey from all dietary supplements. In 2007, the Medicines and Healthcare Products Regulatory Agency (MHRA) of the United Kingdom advised all herbal interest groups to withdraw all unlicensed proprietary products that may contain hepatotoxic PAs from Senecio species. PAs are thus markers for detection of hepatotoxic components in herbs.

General Components Coupled with "Fingerprints"

General components are common and specific components present in a particular species, genus or family. These components may be used with "fingerprints" for quality control purposes.

Lobetyolin, a polyacetylene compound, is used as a marker for Radix Codonopsis (Dangshen) in thin-layer chromatography (TLC). Radix Codonopsis is derived from the roots of three Codonopsis species, namely Codonopsis pilosula (Franch.) Nannf., C. pilosula Nannf. var. modesta (Nannf.) L. T. Shen or C. tangshen Oliv. Our study shows that other five Codonopsis species that are common substitutes of Radix Codonopsis also contain lobetyolin. They are C. tubulosa Kom., C. subglobosa W. W. Smith, C. clematidea (Schynek) C. B. Cl., C. canescens Nannf. and C. lanceolata (Sieb. et Zucc.) Trautv. Moreover, the roots of Campanumoea javanica Bl. and Platycodon grandiflorum (Jacq.) A. DC. (Family Campanulaceae), which are easily confused with Radix Codonopsis, also contain lobetyolin. Therefore, lobetyolin may be used as a general chemical marker coupled with HPLC-UV "fingerprints" to differentiate Radix Codonopsis from its substitutes and adulterants.

As a chemical component may have more than one attribute, a component may belong to multiple categories. For example, ginkgolides A, B and C, and bilobalide are not only characteristic components, but also bioactive components of Ginkgo biloba. Ginsenoside Rg1, Re and Rb1 are both main and bioactive components of Panax ginseng.

In this section, we describe cases to exemplify how chemical markers are used to evaluate the quality of herbal medicines in manufacturing, and as potential lead compounds for new drug development.

Identification of Adulterants

Derived from the resin of Garcinia hanburyi Hook f. (family Guttiferae), gamboges (Tenghuang) has been used in China to treat scabies, tinea and malignant boil, and in Thailand to treat infected wounds, pain and oedema. Characteristic polyprenylated caged xanthones including gambogic acid, gambogenic acid were isolated as the main and bioactive components of gamboges. In our previous study, an adulterant of gamboges was differentiated from the authentic sample by an HPLC-UV method using eight caged xanthones as chemical markers. The chromatogram of gamboges had all eight compounds, while the adulterant showed none of them.

Differentiation of Herbal Medicines with Multiple Sources

Radix Stemonae (Baibu) is a traditional antitussive and insecticidal herbal medicine derived from the roots of three Stemonae species, namely Stemona tuberosa Lour, S. sessilifolia (Miq.) Miq. and S. japonica Miq. The Stemona alkaloids were pharmacologically proven to be responsible for the antitussive and insecticidal effects of Radix Stemonae. In our studies, we observed that the chemical profiles of these three

Stemona species varied greatly. Croomine-typealkaloids such as croomine were detected in all three species, while protostemonine-type alkloids such as protostemonine and maistemonine were detected in S. japonica and S. sessilifolia. Moreover, stichoneurine-type alkaloids such as stemoninine, neotuberostemonine and tuberostemonine were only found in S. tuberosa. Stemona alkaloids may be used as markers to discriminate the three Stemona species.

Determination of the Best Harvesting Time

Rhizoma Chuanxiong (Chuanxiong) is one of the traditional Chinese medicinal herbs frequently used to treat cerebro- and cardio-vascular diseases. Various chemical compounds have been isolated and identified from Rhizoma Chuanxiong, including ferulic acid, senkyunolide I, senkyunolide H, senkyunolide A, coniferyl ferulate, Z-ligustilide, 3-butylidenephthalide, riligustilide and levistolide A. These chemicals have multiple biological activities which may contribute to the therapeutic effects of the herb. Thus, major bioactive components senkyunolide A, coniferyl ferulate, Z-ligustilide, ferulic acid, 3-butylidenephthalide, riligustilide and levistolide A may be used as markers to select the best harvesting time. A previous study using these markers suggested that the best harvesting time for Rhizoma Chuanxiong is from mid April to late May.

Confirmation of Collection Sites

In our studies on the chemistry and antitussive activities of Radix Stemonae, four chemical profiles of S. tuberosa of different geographic sources were characterised using croomine, stemoninine, neotuberostemonine or tuberostemonine as markers. Moreover, the total alkaloid of S. tuberosa exhibited various levels of antitussive activities in a citric acid-induced guinea pig cough model. Croomine, stemoninine, neotuberostemonine and tuberostemonine all possess significant antitussive activities, however, croomine (croomine type) acted on the central nervous system pathway, whereas the other three alkaloids (stichoneurine type) acted on the peripheral pathway of cough reflex. In terms of safety, those containing stichoneurine-type alkaloids are more suitable Radix Stemonae sources than those containing croomine as the major component. Croomine, stemoninine, neotuberostemonine, and tuberostemonine may be used as markers to confirm the collection sites for S. tuberosa (e.g. Shizhu and Erbian in Sichuan Province, Masupo and Baoshan in Yunnan Province, Shanglin in Guangxi Zhuang Autonomous Region or Yudu in Jiangxi Province) which contains higher levels of stemoninine, neotuberostemonine or tuberostemonine, and a low level of croomine.

https://cmjournal.biomedcentral.com/articles/10.1186/1749-8546-3-7

Vocabulary

bioactivity 英 [biːəˈʊæktɪvɪtɪ] 美 [biːəˈʊæktɪvɪtɪ] n. 生物活性；生物活度(指杀虫剂等对生物体的影响)

differentiation 英 [ˌdɪfəˌrenʃɪˈeɪʃn] 美 [ˌdɪfəˌrenʃɪˈeʃən] n. 区别；分化
aerial 英 [ˈeərɪəl] 美 [ˈerɪəl] adj. 空气的；航空的；空中的
correlative components   相关成分
psoralen 英 [ˈsɔːrələn] 美 [ˈsɒrələn] n. 补骨脂素
isopsoralen 英 [aɪsɒpsəˈrɑːlen] 美 [aɪsɒpsəˈrɑːlen] n. 当归根素；白芷素；异补骨脂内酯
psoralenoside   n. 补骨
isopsoralenoside   n. 异补骨脂苷
relieve 英 [rɪˈliːv] 美 [rɪˈliv] vt. 缓解
arthritis 英 [ɑːˈθraɪtɪs] 美 [ɑːrˈθraɪtɪs] n. 关节炎
nephrotoxic 英 [nefrəˈtɒksɪk] 美 [nefrəˈtɒksɪk] adj. 对肾脏有害处的；足以危害肾脏的
hepatic veno-occlusive disease   肝静脉阻塞病
dietary 英 [ˈdaɪətərɪ] 美 [ˈdaɪɪˌterɪ] adj. 饮食的；规定食物的
hepatotoxic 英 [ˌhepətəʊˈtɒksɪk] 美 [ˌhepətoʊˈtɒksɪk] adj. 肝毒素的
stemoninine 英 [stɪmənɪˈnaɪn] 美 [stɪmənɪˈnaɪn] n. 百部新碱
neotuberostemonine   n. 新对叶百部碱
tuberostemonine 英 [tjuːbəˈrɒstiːmənɪn] 美 [tjuːbəˈrɒstiːmənɪn] n. 对叶百部碱；蔓生百部碱
croomine 英 [kˈruːmaɪn] 美 [kˈruːmaɪn] n. 金刚大碱

Useful Expressions

1. document 英 [ˈdɒkjumənt] 美 [ˈdɑːkjumənt] vt. 记录；证明；为……提供证明
He wrote a book documenting his prison experiences.
他写书详细记录了他的狱中经历。

2. moisten 英 [ˈmɔɪsn] 美 [ˈmɔɪsən] vt.（使）变得潮湿；变得湿润
She took a sip of water to moisten her dry throat.
她抿了一口水,润一下发干的喉咙。

3. couple 英 [ˈkʌpl] 美 [ˈkʌpəl] vt. 连在一起；连接
Its engine is coupled to a semiautomatic gearbox.
它的发动机与一个半自动变速箱相连接。

4. discriminate 英 [dɪˈskrɪmɪneɪt] 美 [dɪˈskrɪməˌnet] vt. 歧视；区别；辨出
He is incapable of discriminating between a good idea and a terrible one.
他分不清主意的好坏。

5. confirmation 英 [ˌkɒnfəˈmeɪʃn] 美 [ˌkɑːnfərˈmeɪʃn] n. 证实；证明；确认
We would appreciate confirmation of your refusal of our invitation to take part.
若您确认拒绝我们的邀请、不能参加,我们将感激不尽。

Questions

1. What is main component of herbal medicine?
2. What is correlative component in herbal medicines?
3. What are the toxic components mentioned in the passage?
4. Can you tell us how to identify adulterants?
5. When is the best harvesting time?

4.2 Brief Introduction to Chemical Constituents of Medicinal Plants Resources

Passage 1

Evaluation of Chemical Constituents and Important Mechanism of Pharmacological Biology in Dendrobium Plants (Part 1)

Introduction

The origin of orchids (Orchidaceae) probably has to be backdated to 120 million years ago. It is the largest family group among angiosperms, as well as the most highly evolved family of the flowering plants, with approximately 25,000 to 35,000 species under 750 to 900 genera. Dendrobium species, commonly known as "Shihu" or "Huangcao", is the second largest genus in Orchidaceae. Most Dendrobium species grow best in relatively high and mountainous areas, at 1,400—1,600 m above sea level, at a mild temperature, and in a humid and foggy environment. Characterized by a broad geographical distribution, which allows Dendrobium species to grow into tremendous diversities producing a large number of interspecific hybrids with different morphological features, they are widely distributed in Asia, Australia, and Europe, for instance, in India, Sri Lanka, China, Japan, Korea, New Guinea, and New Caledonia. In the early age, using molecular approaches would delimit the subtribe-Dendrobiinae, affecting approximately 1,100 species (900 in Dendrobium Sw.) from Indo-Asian and Pacific region, and more than 1,200 species in Australasia are from various Dendrobium species. Today, Indo-Asian and Pacific regions have one of the largest and most diverse and taxonomically problematic orchid groups. Interestingly, there has been a long history of the usage of first-rate herbs and folk traditional herbs in India and China.

Since the ancient times (600 BC) in India, the oldest references regarding the use of medicinal herbs are found in the Sanskrit literature, namely "Charaka Samhita". The earliest treatise on "Ayurveda" describes the property of plant drugs and their uses. In the Ayurvedic system of medicine, Flickingeria macraei is used in "Ayurveda". It is commonly named as "jeevanti" and is used as an astringent to the bowels, as an aphrodisiac, and in asthma and bronchitis. Other commonly used orchid drugs in the Ayurvedic system are salem, including jewanti (Dendrobium alpestre). Similarly, it is the first time for China to regard orchids as herbal medicines. The Emperor Shen-Nung advised on the medicinal properties of Dendrobium species in "materia medica" in the 28th century BC. As early as 200 BC, the Chinese pharmacopoeia *The Sang Nung Pen Tsao Ching* mentioned Dendrobium as a source of tonic, astringent, analgesic, and anti-inflammatory substances. In Song Dynasty (960—1279). It mentioned the medicinal uses of orchids, namely Dendrobium species according to "A Diagnosis of Medical Herbs" from the *Zheng Lei Ben Cao*. In Ming Dynasty (1368—1644) many references on the use of orchids as medicinal herbs were available.

The Chemical Compounds of Dendrobium Species

The stem of Dendrobium species has been used in traditional Chinese medicine as a tonic and antipyretic since ancient days for treating human disorders. However, misidentification and adulteration led to a loss of therapeutic potency and potential intoxication. For decades, fast developing molecular techniques using DNA fingerprinting, DNA sequencing, and DNA microarray have been applied extensively to authenticate Chinese medicinal materials, including various Dendrobium species, namely D. aphyllum, D. candidum, D. chrysanthum, D. densiflorum, D. huoshanense, D. gratiosissimum, D. longicornu, D. nobile, D. secundum, D. chrysotoxum, D. crystallinum, D. fimbriatum, and others. In the early years, Williams and Harborne conducted a major survey of leaf flavonoids at the Plant Science Laboratories of the University of Reading in UK. They conducted research on 142 species belonging to 75 genera and found that the most common constituents were flavone C-glycoside and flavonols. Since then, about 100 compounds from 42 Dendrobium species including 32 alkaloids, 6 coumarins, 15 bibenzyls, 4 fluorenones, 22 phenanthrenes, and 7 sesquiterpenoids constituents have been discussed and reviewed. To date, various Dendrobium species are known to produce a variety of secondary metabolites. The biological activities and pharmacological actions of all of these compounds were investigated in detail.

The Pharmacological Effects of Dendrobium Species

1) Antioxidant Activities

Oxidative stress is induced by free radicals that participate in a variety of chemical reactions with biomolecules, leading to a pathological condition. Reactive oxygen species (ROS) is one of the major free radicals. It mainly comprises superoxide (O_2^-) and nitric oxide (NO) radicals, including (i) catalase (CAT), (ii) peroxidase (POD), (iii) ascorbate peroxidase (APX), (iv) 2,2'-azino-bis (3-ethylbenzothiazoline-6-sulphonic acid) or ABTS, (v) hydroxyl, and (vi) 1,1-diphenyl-2-picrylhydrazyl (DPPH) radicals. During these processes, it was found that the content of malondialdehyde (MDA) was increased.

However, cells possess two distinctive antioxidant defense systems to counteract the damage, including enzymatic antioxidants and nonenzymatic antioxidants. Enzymatic antioxidants comprise catalase, superoxide dismutase (SOD), glutathione peroxidase (GPx), and others associated with enzymes/molecules. Nonenzymatic antioxidants include ascorbic acid (vitamin C), α-tocopherol (vitamin E), and β-carotene, and play a key role in removing reactive oxygen species.

Oxidative stresses have been classified as exogenous factors and endogenous factors. Factors related to environmental stress include water/soil drought stress, chilling injury stress, and sound wave stress. There were studies on the effects of cold storage on cut Dendrobium inflorescences, showing that chilling injury symptoms for floral buds and open flowers of Dendrobium could lead to oxidative stress, stemming from the production of reactive oxygen species. In this process, the decrease of cellular functions by peroxidation of membrane lipids when lipoxygenase enzymes are degraded in cells. However, this physiological process in floral buds and open flowers is part of antioxidant defense systems, which can decrease the content of oxygen free radicals and other oxygen compounds. It indicates the protective cellular functions and antioxidant capacity. This is an adaptive nature to ensure some plants survive in the freezing winter.

In addition, sound wave stress is one of the environmental stresses, similar to the low temperature stress. Studies of the effects of sound wave stress on D. candidum indicated that it was a lipid peroxidation parameter which can regulate the levels of MDA. Firstly, the MDA content in different organs is increased and then declines afterwards, which will then be followed by an increase again. That gives a net increase of the level of MDA. It was noted that the MDA level appears to be the lowest when the activities of antioxidant enzymes are the highest. However, the levels of MDA are yet to be fully understood. We believe the antioxidant enzymes could protect plant cells from oxidative damage in sound wave stress. However, the correlation between the mechanisms of antioxidant action and the sound wave stress and chilling injury stress has yet to be identified.

Water deficit caused by soil drought is one of the most frequent environmental stresses. The effects of exogenous drought menace, prompt further increase in the activity of ROS and decrease in the malondialdehyde (MDA) content, to prevent the breakage of DNA and protein degradation from doing damage to plant life. It is well known that nitric oxide (NO) is a ubiquitous signal molecule involved in many life processes of plants, seed germination, hypocotyl elongation, leaf, expansion, root growth, lateral roots initiation, and apoptosis, and so forth. It is also involved in multiple plant responses to environmental stress. Data demonstrated that, at lower concentrations of exogenous NO with 50 mmol L-1 SNP, the activation of POD, SOD, and CAT was significantly increased, and the MDA content decreased with 50Ml SNP. NO could protect D. huoshanense against the oxidative insult caused by a drought stress; meanwhile, a high level of RWC can be maintained. Furthermore, it suggests that the molecular messenger NO can trigger epigenetic variation and increase the demethylation ratio of methylated sites for D. huoshanense by using DNA analysis. These results may imply that the expression of some genes is involved in the response to drought stress triggered by NO.

Numerous Dendrobium species such as D. nobile, D. denneanum, D. huoshanense, D. chrysotoxum, D. moniliforme, D. tosaense, D. linawianum, D. candidum, D. loddigesii, and D. fimbriatum have polysaccharides as the active compounds. Polysaccharides play important roles in many biological processes and are used to treat various diseases. Polysaccharides isolated from D. nobile, D. huoshanense, D. chrysotoxum, and D. fimbriatum species manifest antioxidant and free radical scavenging activities. D. nobile polysaccharide displayed the highest scavenging activity toward hydroxyl, ABTS, and DPPH free radicals. However, D. fimbriatum polysaccharide has significant scavenging action toward ABTS free radicals, and it also shows a weak DPPH free radical scavenging action. On the contrary, D. denneanum polysaccharide exerts a powerful DPPH free radical scavenging action, but its ABTS scavenging effect is not obvious. In addition, both D. huoshanense and D. chrysotoxum polysaccharides reveal an insufficient ABTS free radical scavenging effect. However, it manifests potential hydroxyl radical scavenging activity. Overall, the free radical scavenging activities of polysaccharides on hydroxyl and DPPH free radicals remain at a high level, but the inhibitory effect on ABTS free radicals is weak. The polysaccharide of D. huoshanense has a backbone of $(1 \rightarrow 4)$-linked α-D-Glcp, $(1 \rightarrow 6)$-linked α-D-Glcp, and $(1 \rightarrow 4)$-linked β-D-Manp from mannose (Man), glucose (Glc), and a trace of galactose (Gal). Its antioxidant effect in the livers of CCl4-treated mice was evidenced by a decrease of malondialdehyde (MDA) and increase of superoxide dismutase (SOD), catalase (CAT), and glutathione peroxidase (GPx).

https://www.hindawi.com/journals/ecam/2015/841752/

Vocabulary

dendrobium 英 [dend'rəʊbjəm] 美 [dend'roʊbjəm] *n*. 石斛兰
Dendrobium species   石斛兰属
Orchidaceae 英 [ɔːˈkɪdəsiˌiː] 美 [ɔːˈkɪdəsiˌiː] *n*. [植] 兰科
interspecific hybrid   种间杂种
aphrodisiac 英 [ˌæfrəˈdɪziæk] 美 [ˌæfrəˈdɪziˌæk, -ˈdizi-] *n*. 壮阳剂
asthma 英 [ˈæsmə] 美 [ˈæzmə] *n*. [医] 气喘；哮喘
bronchitis 英 [brɒŋˈkaɪtɪs] 美 [brɑːŋˈkaɪtɪs] *n*. 支气管炎
analgesic 英 [ˌænəlˈdʒiːzɪk] 美 [ˌænəlˈdʒizɪk, -sɪk] *n*. 止痛剂；镇痛剂
antipyretic 英 [ˌæntɪpaɪˈretɪk] 美 [ˌæntɪpaɪˈretɪk, ˌæntaɪ-] *adj*. 退热的；退烧的
adulteration 英 [əˌdʌltəˈreɪʃn] 美 [əˌdʌltəˈreɪʃn] *n*. 掺假；掺杂；掺假货

potency 英 [ˈpəʊtnsi] 美 [ˈpoʊtnsi] n. 效力；潜能；权势；(男人的)性交能力
authenticate 英 [ɔːˈθentɪkeɪt] 美 [ɔˈθentɪˌket] vt. 证明是真实的、可靠的或有效的
metabolite 英 [meˈtæbəlaɪt] 美 [meˈtæbəlaɪt] n. 代谢物
oxidative stress 氧化应激
biomolecules 英 [baɪɒməʊˈlekjuːl] 美 [baɪɒmoʊˈlekjuːl] n. 生物分子
pathological 英 [ˌpæθəˈlɒdʒɪkl] 美 [ˌpæθəˈlɑːdʒɪkl] adj. [医] 病理学的；疾病的
enzymatic antioxidant 酶抗氧剂
nonenzymatic antioxidant 非酶抗氧剂
reactive oxygen species 活性氧
exogenous 英 [ekˈsɒdʒənəs] 美 [ekˈsɑːdʒənəs] adj. 外生的；外成的；外因的
endogenous 英 [enˈdɒdʒənəs] 美 [enˈdɑːdʒənəs] adj. 内长的；内生的
cellular 英 [ˈseljələ(r)] 美 [ˈsɛljələ·] adj. 细胞的；蜂窝状的；由细胞组成的
exogenous drought menace 外源性干旱威胁
degradation 英 [ˌdegrəˈdeɪʃn] 美 [ˌdɛgrəˈdeʃən] n. 恶化；堕落；潦倒；毁坏
scavenge 英 [ˈskævɪndʒ] 美 [ˈskævəndʒ] vt. 清除污物；(在废物中)寻觅

Useful Expressions

1. orchid 英 [ˈɔːkɪd] 美 [ˈɔːrkɪd] n. 兰花
Vanilla is extracted from a tropical orchid.
香草香精是从一种热带兰提制出来的。
2. evolve 英 [iˈvɒlv] 美 [iˈvɑːlv] vi. 发展；[生] 通过进化进程发展或发生
The bright plumage of many male birds has evolved to attract females.
很多雄鸟进化出鲜艳的羽毛是为了吸引雌鸟。
3. in detail 详细地
We examine the wording in detail before deciding on the final text.
我们彻底地检查了措辞后才最终定稿。
4. deficit 英 [ˈdefɪsɪt] 美 [ˈdɛfɪsɪt] n. 赤字；亏损；亏空；不足额
They're ready to cut the federal budget deficit for the next fiscal year.
他们已准备好削减下一财年的联邦预算赤字。

Questions

1. Where are orchids (Orchidaceae) distributed?
2. How many chemical compounds of Dendrobium species are mentioned in this passage?
3. What pharmacological effects do Dendrobium species have?

Passage 2

Evaluation of Chemical Constituents and Important Mechanism of Pharmacological Biology in Dendrobium Plants (Part 2)

The Antioxidant Activities of Bibenzyl Derivatives, Phenanthrenes, and Stilbenes from D. candidum and D. loddigesii

Four new bibenzyl derivatives, dendrocandin F, dendrocandin G, dendrocandin H, and dendrocandin I, are extracted from D. candidum. They act as antioxidant agents to clear up the free radicals. Accordingly, the IC_{50} values of dendrocandin F and dendrocandin G are 55.8 mM and 32.4 mM, respectively. A series of structurally related compounds from D. loddigesii, known as loddigesiinol A—D, suppress the production of nitric oxide (NO) with IC_{50} values of 2.6mM for loddigesiinol A, 10.9mM for loddigesiinol B, and 69.7mM for loddigesiinol D. Further study indicated that phenanthrenes and dihydrophenanthrenesin D. loddigesii have significant effects on inhibiting the production of NO and DPPH free radicals. These compounds are known as (numbered as 1a—7a) phenanthrenes (1a), phenanthrenes (2a), phenanthrenes (3a), phenanthrenes (4a), dihydrophenanthrenes (5a), dihydrophenanthrenes (6a), and dihydrophenanthrenes (7a). Based on the above results, the scavenging activity of loddigesiinols toward NO free radicals is pronounced. The inhibitory actions of phenanthrenes and dihydrophenanthrenes from D. loddigesii on the production of NO and DPPH free radicals are recognised.

Anti-Inflammatory Activity

1) Inhibition of Nitric Oxide (NO) by D. nobile and D. chrysanthum Constituents

It is known that, upon lipopolysaccharide (LPS) stimulation, macrophages produce a large amount of inflammatory factors, such as tumor necrosis factor α (TNF-α), interleukin-1 beta (IL-1β), interleukin 6 (IL-6), interferon (IFN), and other cytokines. LPS induces endogenous nitric oxide (NO) biosynthesis through induction of inducible nitric oxide synthase (iNOS) in macrophages, which is involved in inflammatory responses. D. nobile and D. chrysanthum constituents strongly inhibit TNF-α, IL-1β, IL-6, and the NO production.

2) D. nobile Constituents

From D. nobile two new bibenzyl derivatives, namely nobilin D (given number 1) and nobilin E (given number 2), and a new fluorenone, namely nobilone (given number 3), together with seven known compounds (given numbers 4—10), including $R_1=R_2=OCH_3$ $R_3=OH$ $R_4=H$ (4), $R_1=R_2=R_3=OCH_3$ $R_4=H$ (5), $R_1=R_2=OH$ $R_3=R_4=H$ (6), $R_1=R_3=OCH_3$ $R_2=OH$ $R_4=H$ (7), $R_1=OCH_3$ $R_2=R_3=OH$ $R_4=H$ (8), $R_1=R_3=OH$

$R_2=R_4=H$ (9), and $R_1=R_3=OH$ $R_2=HR_4=OCH_3$ (10), have been identified. Compounds 1—3, 5, and 8—10 can inhibit NO production. On the other hand, compounds 2 and 10 can exhibit a strong to resveratrol, while enhanced cytotoxic potential has been found in compounds 4 and 7. Compounds 1—10 act as NO inhibitors, as evidenced by oxygen radical absorbency capacity (ORAC) assays. These findings are focused on all of these compounds, except compound 6, and their inhibitory effects on NO production, in murine macrophages (RAW 264.7), are activated by LPS and interferon (IFN-ç). Furthermore, it was found that a new phenanthrene, together with nine known phenanthrenes and three known bibenzyls, manifests an inhibitory effect on NO production in RAW 264.7 cells. The structures of all of the compounds (given numbers A—M), including $R_1=R_3=OH$, $R_2=R_5=OCH_3$, $R_4=H(A)$, $R_1=R_4=R_5=OH$, $R_2=H$, $R_3=OCH_3(B)$, $R_1=R_5=OH$, $R_2=R_3=OCH_3$, $R_4=H(C)$, $R_1=R_2=OCH_3$, $R_3=R_5=OH$, $R_4=H(D)$, $R_1=H$ $R_2=R_3=OH(E)$, $R_1=H$, $R_2=OH$, $R_3=OCH_3(F)$, $R_1=OCH_3$, $R_2=R_4=OH$, $R_3=R_5=H(G)$, $R_1=R_3=OH$, $R_2=OCH_3(H)$, $R_1=R_5=OH$, $R_2=R_4=H$, $R_3=OCH_3(I)$, $R_1=R_3=OCH_3$, $R_2=R_5=OH$, $R_4=H(J)$, $R_1=R_2=R_4=OH$, $R_3=OCH_3$, $R_5=H(K)$, $R_1=OCH_3$, $R_2=R_4=H$, $R_3=R_5=OH(L)$, and $R_1=R_2=R_3=OH$, $R_4=R_5=H(M)$, were elucidated by analysis of the spectroscopic data, including extensive two-dimensional nuclear magnetic resonance spectroscopy (2D-NMR) and mass spectrometry (MS). All of compounds C, D, I, K, and L are relatively strong inhibitors of NO production, and their potencies are better than those of compounds A and E—H. Compounds C, D, I, K, and L could inhibit NO synthesis which might contribute to the suppression of LPS-induced TNF-α and IL-1β production.

3) D. chrysanthum Constituents

Anti-inflammatory compounds from ethyl acetate extracts of D. chrysanthum species were evaluated. These structures were elucidated on the basis of the high-resolution mass spectrometry, NMR spectroscopy, and X-ray crystal diffraction analysis. A novel phenanthrene phenol derivative with a spironolactone ring (dendrochrysanene), namely 2-hydro-7,70,100-trihydroxy-4-40-dimethoxylspiro[(1H)-cyclopenta[a]naphthalene-3,30-(20H)-phenanthro[2,1-b]furan]-1,20-dione, was identified. Results showed anti-inflammatory activity of dendrochrysanene on iNOS mRNA induced by LPS in mouse peritoneal macrophages. Meanwhile, there are several inflammatory cytokines, such as TNF-α, IL-8, and IL-10, which were inhibited. Thus, the beneficial effects of dendrochrysanene compounds as anti-inflammatory agents were identified.

Sjögren's Syndrome

Sjögren's syndrome (SS) is a chronic autoimmune disease with disorder of the exocrine glands. The symptoms are dry eyes, dry mouth, dry throat and thirst with blurred vision, and so forth. The abnormal distribution of salivary glands in SS leads to an inflammatory effect and pathological processes triggering lymphocyte infiltration and apoptosis. Lymphocyte infiltration and apoptotic pathways, shown by regulation of Bax, Bcl-2, and caspase-3, have been reported in submandibular glands (SG). Furthermore, in pathological processes were also found that the subsequent signal of cell expression to mRNA of proinflammatory cytokines, such 30 as tumor necrosis (TNF-a), interleukin (IL-1β), and IL-6. As well as, a high level of expression of aquaporin-5 (AQP-5) is identified. AQP-5 is one of a water channel protein, and plays an important role in 31 the salivary secretions.

Treatment with D. officinale polysaccharides (DP) could suppress the progression of lymphocyte infiltration and apoptosis in SS and rectify the chaos of proinflammatory cytokines including TNF-α, IL-1 beta, and IL-6 in SG. mRNA expression of TNF-α was inhibited. The process of regulation resulted in a series of marked responses, such as translocation of NF-κB, prolonged MAPK, cytochrome C release, and caspase-3 activation, which have been identified. In addition to the findings on DP treatment, there are reports on the findings on the extracts of D. candidum and D. nobile.

Chrysotoxine isolated from D. nobile can increase the expression of AQP-5 in dry eyes, one of the symptoms of SS, and restore the distribution of AQP-5 in lacrimal glands and corneal epithelia, by inhibiting the release of cytokines, such as IL-1, IL-6, and TNF-α. In the meantime, it can activate the mitogen-activated protein kinase (MAPK) signaling pathways. Furthermore, it elicits an increase in the production of matrix metalloproteinase-9 (MMP-9). Ultimately, it leads to an increase of saliva and tear secretion. The medical efficacy of the extracts of D. officinale, D. nobile, and D. candidum species has been demonstrated.

Neuroprotective Effect (Parkinsonian Syndrome)

Five bioactive derivatives isolated from Dendrobium species, namely chrysotoxine (CTX), moscatilin, crepidatin, nobilin B, and chrysotobibenzyl, are potential neuroprotective compounds with antioxidant activity which can be used in the treatment of Parkinson's disease (PD). The IC_{50} values of DPPH free radical scavenging activities of chrysotoxine (CTX), moscatilin, crepidatin, and nobilin B were 20.8 ± 0.9, 28.1 ± 7.1, 38.2 ± 3.5, and 22.2 ± 1.4, respectively. The above data show that crepidatin is better than other compounds in DPPH free radical scavenging capacity. CTX could selectively antagonize MPP+ in dopaminergic pathways in the brain; also 6-hydroxydopamine (6-OHDA) has been inhibited by mitochondrial protection and NF-κB modulation in SH-SY5Y cells. Thus, it could explain how it may be beneficial in preventing PD.

In addition, results of CTX compound in Dendrobium nobile Lindl. used in treatment of PD have also been clearly reported. We evaluated the pharmacokinetics of oral (100 mg/kg) and intravenous (25 mg/kg) administration of CTX preparation using high performance liquid chromatography-tandem mass spectrometric (HPLC-MS/MS) method in animal tests. Results indicate efficacy and safety in treating PD conditions. The antioxidant mechanisms towards 6-OHDA and SH-SY5Y and regulation of antiapoptosis in cell signaling pathways were described.

https://www.hindawi.com/journals/ecam/2015/841752/

Vocabulary

  antioxidant activity 抗氧化剂活性
  bibenzyl derivative 联苯苄衍生物
  dendrobium nobile 金钗石斛
  Dendrobium chrysanthum 束花石斛
  salivary secretion 唾液分泌
  D. candidum 铁皮石斛
  neuroprotective effect 神经保护作用
  pharmacokinetics 英 [fɑːməkəʊkaɪˈnetɪks] 美 [ˌfɑːmɪkoʊkˈnetɪks] *n.* 药物（代谢）动力学

Useful Expressions

1. oral 英 [ˈɔːrəl] 美 [ˈɔrəl] *adj.* 口服的；属于或关于嘴的
Painful typhoid injections are a thing of the past, thanks to the introduction of an oral vaccine.
自从发明了口服疫苗后,令人痛苦的伤寒疫苗注射就成了过去。

2. intravenous 英 [ˌɪntrəˈviːnəs] 美 [ˌɪntrəˈvinəs] *adj.* 静脉注射的
Objective: To determine the effect of fentanyl intravenous analgesia on plasma motilin levels.
目的:研究芬太尼静脉镇痛对血浆胃动素水平的影响。

Questions

1. What are the functions of extracts from D. candidum?
2. What constituents do D. nobile and D. chrysanthum have?
3. What are the five bioactive derivatives isolated from Dendrobium species?

4.3 Main Biosynthetic Pathway and Research Examples of Chemical Constituents of Medicinal Plants Resources

Passage 1

Exploring Drug Targets in Isoprenoid Biosynthetic Pathway for Plasmodium Falciparum (Extract)

Antimalarial Compounds Against Isoprenoid Biosynthetic Pathway

Biosynthesis of several isoprenoids in P. falciparum was studied and terpenes (molecules with a similar chemical structure to the intermediates of the isoprenoids pathway) as potential antimalarial drugs were evaluated. Different terpenes and S-farnesylthiosalicylic acid were tested on cultures of the intraerythrocytic stages of P. falciparum, and the 50% inhibitory concentration for each one was found. Farnesol, nerolidol, limonene, and linalool terpenes have been used against P. falciparum. All the terpenes tested inhibited dolichol biosynthesis in the trophozoite and schizont stages. Farnesol, nerolidol, and linalool showed stronger inhibitory activity in the biosynthesis of the isoprenic side chain of the benzoquinone ring of ubiquinones in the schizont stage. The inhibitory effect of terpenes and S-farnesylthiosalicylic acid on the biosynthesis of both dolichol, the isoprenic side chain of ubiquinones, and the isoprenylation of proteins in the intraerythrocytic stages of P. falciparum appears to be specific, because overall protein biosynthesis was not affected.

A variety of proteins undergo posttranslational modification such as prenylation near the carboxyl terminus with farnesyl (C15) or geranylgeranyl (C20) groups. Protein farnesyltransferase (PFT) transfers the farnesyl group from farnesyl diphosphate to the SH of the cysteine near the C-terminus of proteins such as Ras. PFT inhibitors (PFTIs) have been extensively developed as anticancer agents because of their ability to block tumor growth in experimental animals. PFTIs have been explored as antimalarial and antitrypanosome agents because these compounds are much more toxic to these parasites than to mammalian cells, and there are a large number of lead compounds which have been explored as part of the antiparasite drug discovery program. MEP pathway inhibition with fosmidomycin reduces protein prenylation, confirming that de novo isoprenoid biosynthesis produces the isoprenyl substrates for protein prenylation. One important group of prenylated proteins is small GTPases, such as Rab family members which mediate cellular vesicular trafficking.

Substituted tetrahydroquinolines (THQs) have been previously identified as inhibitors of mammalian protein farnesyltransferase. Fletcher et al. designed and synthesized a series of inhibitors that are selective for P. falciparum farnesyltransferase (PfPFT) in 2008. Several PfPFT inhibitors have been found to inhibit the malarial enzyme with IC_{50} values down to 1nM, and that blocks the growth of P. falciparum in infected whole cells (erythrocytes) with ED50 values down to 55nM. Potent, Plasmodium-selective farnesyltransferase inhibitors that arrest the growth of malaria parasites have been explored. A new synthetic pathway was devised to reach tetrasubstituted 3-arylthiophene 2-carboxylic acids in a three-step solid-phase synthesis. This very efficient methodology provided more than 20 new compounds that were evaluated for their ability to inhibit protein farnesyltransferase from different species as well as Trypanosoma brucei and P. falciparum proliferation.

Strong inhibition of P. falciparum PFT (PfPFT) by peptidomimetics illustrated the potential of targeting these enzymes in developing drug therapy for malaria. Ohkanda et al. have recently demonstrated the potency of a variety of other peptidomimetics as inhibitors of P. falciparum growth and PfPFT activity. Moura et al. have also shown that the monoterpene, limonene, inhibits parasite development and prenylation of P. falciparum proteins. The enzymes of the nonmevalonate pathway for isoprenoid biosynthesis are attractive targets for the development of novel drugs against malaria and tuberculosis.

Fosmidomycin and FR900098 (an N-acetyl derivative of fosmidomycin) are inhibitor of DOXP reductoisomerase, which show antimalarial activity *in vitro* and *in vivo*.

Reverse hydroxamate-based inhibitor for IspC enzyme was evaluated by Behrendt et al. Fosmidomycin has been proven to be efficient in the treatment of P. falciparum malaria by inhibiting 1-deoxy-D-xylulose-5-phosphate reductoisomerase (DXR), an enzyme of the nonmevalonate pathway, which is absent in humans.

DXR represents an essential enzyme of the mevalonate-independent pathway of the isoprenoid biosynthesis. Using fosmidomycin as a specific inhibitor of DXR, this enzyme was previously validated as target for the treatment of malaria and bacterial infections. The replacement of the formyl residue of fosmidomycin by spacious acyl residues yielded inhibitors active in the micromolar range. As predicted by flexible docking, evidence was obtained for the formation of a hydrogen bond between an appropriately placed carbonyl group in the acyl residue and the main-chain NH of Met214 located in the flexible catalytic loop of the enzyme. Specific inhibition of enzymes of the nonmevalonate pathway is a promising strategy for the development of novel antiplasmodial drugs. Aryl-substituted-oxa isosteres of fosmidomycin with a reverse orientation of the hydroxamic acid group were synthesized and evaluated for their inhibitory activity against recombinant 1-deoxy-d-xylulose 5-phosphate reductoisomerase (IspC) of Plasmodium falciparum and for their in vitro antiplasmodial activity against chloroquine-sensitive and resistant strains of P.

falciparum. The most active derivative inhibits IspC protein of P. falciparum (PfIspC) with an IC_{50} value of 12nM and shows potent in vitro antiplasmodial activity.

The structure-activity relationships for 43 inhibitors of 1-deoxyxylulose-5-phosphate (DOXP) reductoisomerase, derived from protein-based docking, ligand-based 3D QSAR, and a combination of both approaches as realized by AFMoC (adaptation of fields for molecular comparison) have been presented by Silber et al. A series of novel 3'-amido-3'-deoxy-N(6)-(1-naphthylmethyl) adenosines was synthesized applying a polymer-assisted solution phase (PASP) protocol and was tested for antimalarial activity versus the Dd2 strain of P. falciparum. Further, this series and 62 adenosine derivatives were analyzed regarding 1-deoxy-D-xylulose-5-phosphate reductoisomerase inhibition. Biological evaluations revealed that the investigated 3', N(6)-disubstituted adenosine derivatives displayed moderate but significant activity against the P. falciparum parasite in the low-micromolar range.

Genetic Polymorphism in P. falciparum

P. falciparum malaria is an example of evolutionary selection. Both host and parasite show the phenomenon of natural selection. Several polymorphisms in human host and parasite have been found. In the drug discovery process the polymorphisms in parasite genes encoding the target protein should be considered. In P. falciparum the genetic polymorphisms at 10 loci are considered potential targets for specific antimalarial vaccines. The polymorphism is unevenly distributed among the loci; loci encoding proteins expressed on the surface of the sporozoite or the merozoite (AMA-1, CSP, LSA-1, MSP-1, MSP-2, and MSP-3) are more polymorphic than those expressed during the sexual stages or inside the parasite (EBA-175, Pfs25, PF48/45, and RAP-1). Comparison of synonymous and nonsynonymous substitutions indicates that natural selection may account for the polymorphism observed at seven of the 10 loci studied. This inference depends on the assumption that synonymous substitutions are neutral. The authors obtained evidence for an overall trend towards increasing $A+T$ richness but no evidence of mutation. Although the neutrality of synonymous substitutions is not definitely established, this trend towards an $A+T$ rich genome cannot explain the accumulation of substitutions at least in the case of four genes (AMA-1, CSP, LSA-1, and PF48/45) because the G→C transversions are more frequent than expected. Predicted polymorphisms in genes encoding isoprenoid biosynthesis may play an important role in drug response.

Nonsynonymous single nucleotide polymorphisms in 4-hydroxy-3-methylbut-2-enyl diphosphate reductase (LytB), 2-C-methyl-D-erythritol 4-phosphate cytidylyltransferase, putative (IspD), 1-deoxy-D-xylulose 5-phosphate synthase, 1-deoxy-D-xylulose 5-phosphate reductoisomerase (DOXR), and 4-diphosphocytidyl-2c-methyl-D-erythritol kinase (CMK) proteins may change the amino acids. Such change may affect the proteins and interaction with antimalarial drugs.

1) DNA Sequence Variation(s) and Evolution

Mutations or genetic variations that have ability to alter the activity or availability of transcription factors, as well as mutations that alter the cis-regulatory sequences (transcription factor binding sites) to which they bind, can change expression of gene and both types of changes contribute to evolution. Several studies have suggested that mutations affecting cis-regulatory activity (transcription factor binding site) are the predominant source of expression divergence between species. Changes in gene expression often alter phenotypes; mutations that affect gene expression can affect fitness and contribute to adaptive evolution. DNA sequence variations reported in genes involved in dolichol pathway of P. falciparum were shown.

https://www.hindawi.com/journals/bri/2014/657189/

Vocabulary

Isoprenoid Biosynthetic Pathway　类异戊二烯生物合成途径
terpenes 英 [tɜːˈpiːnɪz] 美 [tɜːˈpiːnɪz] n. 萜烯
biosynthesis 英 [ˌbaɪəʊˈsɪnθɪsɪs] 美 [ˌbaɪoʊˈsɪnθɪsɪs] n. 生物合成
prenylation 英 [prenɪˈleɪʃn] 美 [prenɪˈleɪʃn] n. 异戊烯化
carboxyl terminus 英 [kɑːˈbɒksɪl ˈtɜːmənəs] 美 [karˈbaksəl ˈtɜːmənəs] n. 羧基端
vesicular trafficking　囊泡运输；膜泡运输
mammalian protein farnesyltransferase　哺乳动物蛋白法尼基转移酶
tuberculosis 英 [tjuːˌbɜːkjuˈləʊsɪs] 美 [tuːˌbɜːrkjəˈloʊsɪs] n. 肺结核；[医] 结核病
inhibitory 英 [ɪnˈhɪbɪtəri] 美 [ɪnˈhɪbɪtəri] adj. 禁止的；抑制的
recombinant 英 [rɪˈkɒmbɪnənt] 美 [riːˈkɒmbənənt] n. 重组（复合）器官；重组细胞
polymorphism 英 [ˌpɒlɪˈmɔːfɪzəm] 美 [ˌpɒlɪˈmɔːfɪzəm] n. 多型现象；多态性
P. falciparum　恶性疟原虫；疟原虫
vaccine 英 [ˈvæksiːn] 美 [vækˈsiːn] n. 疫苗；痘苗
synonymous 英 [sɪˈnɒnɪməs] 美 [sɪˈnɑːnɪməs] adj. 同义的；类义的
substitution 英 [ˌsʌbstɪˈtjuːʃn] 美 [ˌsʌbstɪˈtuːʃən, -ˈtjuː-] n. 替换；代替
phenotype 英 [ˈfiːnətaɪp] 美 [ˈfiːnətaɪp] n. 显型

Useful Expressions

1. undergo 英 [ˌʌndəˈgəʊ] 美 [ˌʌndərˈgoʊ] vt. 经历；经验；遭受；承受
He underwent an agonising 48-hour wait for the results of tests.
他苦苦等待了48个小时,化验结果才出来。

2. substitute 英 [ˈsʌbstɪtjuːt] 美 [ˈsʌbstɪtuːt] vt. 代替；替换；代用
They were substituting violence for dialogue.
他们在用暴力取代对话。

3. synthesize 美 [ˈsɪnθɪˌsaɪz] vt. 综合；人工合成；（通过化学手段或生物过程）合成

After extensive research, Albert Hoffman first succeeded in synthesizing the acid in 1938.

经过大量研究,艾伯特·霍夫曼于1938年首次成功合成了该酸性物质。

4. parasite 英 [ˈpærəsaɪt] 美 [ˈpærəˌsaɪt] n. 食客;寄生虫;寄生植物

One of the ways the parasite spreads is through fecal matter.

排泄物是寄生虫传播的一种方式。

5. mediate 英 [ˈmiːdieɪt] 美 [ˈmidiˌet] vt. 经调解解决;斡旋促成

My mom was the one who mediated between Zelda and her mom.

我妈妈充当了泽尔达和她妈妈之间的调解人。

Questions

1. Who synthesized a series of inhibitors that are selective for P. falciparum farnesyltransferase?

2. What does inhibition of P. falciparum PFT (PfPFT) by peptidomimetics show?

3. What is the relationship between DNA Sequence Variation(s) and Evolution according to the passage?

Passage 2

Traditional Uses, Chemical Constituents, and Biological Activities of Bixa Orellana L.: A Review (Extract)

Chemical Compounds

Bixin, a red-colored carotenoid, is the pigment present in high concentration in the annatto seed aril. It is the main substance responsible for the dyeing characteristics of seeds, where its concentration can be as high as 5.0%. However, different seeds may have levels less than 2.0%, and because their commercial value is based on the bixin percentage, levels higher than 2.5% are usually required for export.

Bixin was isolated for the first time from the seeds of Bixa orellana in 1875 and in 1961 its complete chemical structure and stereochemistry were determined by ^{1}H and ^{13}C-NMR. Bixin belongs to the small class of natural apocarotenoids, whose formation occurs by the oxidative degradation of C_{40} carotenoids.

Bixin consists of a chain of 25 carbons and has the molecular formula $C_{25}H_{30}O_4$ (MW = 394.51). It has a carboxylic acid and methyl ester group at the ends of the chain. Bixin occurs in nature as 16-Z (*cis*), but during the extraction process it isomerizes resulting in the 16-E form (*trans*), which is called isobixin.

Many other carotenoids (C_{19}, C_{22}, C_{24}, C_{25}, C_{30}, and C_{32}) occur in Bixa orellana but constitute a minor percentage of the pigments. The major oily constituent of annatto seeds is geranylgeraniol, representing 1% of dry seeds. Norbixin is a demethylated derivative of bixin and although it is a naturally occurring compound, it is almost always referred to as a saponification product of bixin. This is the form used for commercial purposes.

Currently, more than two dozen substances have been isolated from the seeds of Bixa orellana. Besides bixin and norbixin, other compounds such as isobixin, beta-carotene, cryptoxanthin, lutein, zeaxanthin, orellin, bixein, bixol, crocetin, ishwarane, ellagic acid, salicylic acid, threonine, tomentosic acid, tryptophan, and phenylalanine have been found in the seeds of annatto. In addition, the following compounds, in their respective concentrations, are found in these seeds: 40% to 45% cellulose, 3.5% to 5.5% sugars, 0.3% to 0.9% essential oils, 3% fixed oils, 1.0% to 4.5% pigments, and 13% to 16% proteins and alpha and beta-carotene, as well as tannins and saponins.

Mercadante et al. isolated apocarotenoids from annatto seeds: methyl (7Z, 9Z, 9'Z)-apo-6'-lycopenoate, methyl (9Z)-apo-8'-lycopenoate, methyl 1(all-E)-apo-8'-lycopenoate, methyl (all-E)-8-apo-beta-carotene-8'-oate, methyl (all-E)-apo-6'-lycopenoate, 6-geranylgeranyl-8'-methyl-6,8'diapocaroten-6-8'dioate, 6'-geranylgeranyl-6'-methyl-(9Z)-6, 6'-diapocaroten-6-6'-dioate, and 6-geranylgeranyl-6'-methyl-6-6'-diapocaroten-6-6'-dioate.

More than 100 volatile compounds have been detected in aqueous and organic extracts, where 50 of these have already been identified (e.g. bornyl acetate, ∝-caryophyllene, copaene, ∝-cubebene, (+)-cyclosativene, geranyl phenylacetate, 1-heptanetiol, 3-methylpyridine, 4-methylpyridine γ-elemene, β-humulene, isoledene, β-pinene, seline-6-en-4-ol, δ-selinene, (-)-spathulenol, and (+)-ylangene).

Because annatto is a rich source of carotenoids it is of great commercial importance. In fact, the therapeutic properties of annatto (e.g. antioxidant and hypoglycemic) have been attributed to its high levels of carotenoids.

The pigments in annatto seeds can be extracted by mechanical processes through grinding the seeds and by physical-chemical methods using solvents or enzymes. The solvent extraction can be performed by using three basic methods: alkaline extraction (NaOH or KOH solutions), which results in the conversion of bixin to norbixin; extraction with oil (soybean, corn); and extraction using organic solvents (hexane, chloroform, ethanol, acetone, or propylene glycol), which results in the purest form of pigments.

Barbosa-Filho et al. studied the seeds of four types of annatto cultivated in Paraíba State, Brazil, namely, "cascaverde" ("green peel"), "cascavermelha" ("red bark"), "bico de calango" ("lizard beak"), and "graopreto"("black grain"), with respect to their oil (material extracted with hexane) and solid (material extracted with chloroform) contents, and also pure bixin, which was obtained by successive recrystallization from the chloroform fraction. Pure bixin appears as red-purple crystals with a melting point of

196℃—198℃. The different concentrations found for the oil fraction, chloroform extract, and bixin are as follows: red bark 5.8%, lizard beak 5.1%, green peel 4.9%, and black grain 4.6%. Red bark shows the highest yield for both solvent fractions, and the bixin amount is around 1%. This species has been reported as the most used in folk medicine. On the other hand, black grain shows negligible amounts of bixin.

Biological Activity

In 38 studies performed with annatto in 15 different countries in the American countries, to obtain the extracts and fractions tested, several plant parts were used, such as leaf, root, seed, shoot, and even the whole plant. The data surveyed were classified according to the pharmacological activity tested.

Among the twenty-one activities tested, those with the largest number of studies performed were antifungal activity (12), antibacterial activity (12), antimalarial activity (6), and mutagenic activity (3). Cytotoxic activity and toxicity have been little studied, with three and two studies, respectively. Pharmacological activities have been evaluated in animal models (22 preclinical studies), human models (1 clinical study), cell cultures (2 studies), and *in vitro* tests (32 studies).

Antifungal activity has been investigated in one country in Central America (Guatemala) and in two countries in South America (Ecuador and Argentina) using eleven different fungal strains.

Freixa et al. conducted a study in Ecuador to assess the antifungal activity of extracts from the dried leaves of the annatto tree in response to 7 fungi species, obtaining satisfactory antifungal activity against Trichophyton mentagrophytes trains. In Guatemala, three different strains were used to evaluate antifungal activity, with no satisfactory activity being observed.

The extracts of annatto leaves have been evaluated for antibacterial activity against 8 different bacterial strains (Bacillus subtilis, Escherichia coli, Micrococcus luteus, Pseudomonas aeruginosa, Staphylococcus aureus, Salmonella typhi, Shigella dysenteriae, and Staphylococcus epidermidis), showing no activity.

Antimalarial activity has been determined against Plasmodium gallinaceum, Plasmodium lophurae, and Plasmodium berghei. Although the studies conducted previously in the United States did not show significant results, Valdés et al. reported a moderate activity of the seed extracts of Bixa orellana against Plasmodium berghei and falciparum.

Mutagenic and Cytotoxic Activities

No significant effect was observed when extracts of annatto seeds were tested for mutagenic activity in studies performed in the United States and Brazil.

Extracts obtained from annatto seeds and leaves have been tested in cell cultures and the brine shrimp assay, respectively, and have been found to lack cytotoxicity in either model used. These experiments were carried out in Guatemala and the Dominican Republic.

On the other hand, a study performed in Cuba with 10 medicinal plants that were active in inhibiting human lung carcinoma cell growth showed that the ethanolic extract of Bixa orellana presented cytotoxicity at concentrations below 100 ug/mL.

Toxicological Activities

Currently, concerns about the effect of synthetic dyes on human health are incontestable, making people increasingly choose those of natural origin, believing that they are devoid of toxic effects. This is not entirely true, because even a medication from a natural source can be a poison, depending on the dose that is administered. The failure to require in-depth data related to toxicological and chemical analyses for the registration of food additives derived from natural sources certainly makes the information about possible unwanted effects and/or pharmacological activities resulting from their use, much rarer than expected in view of the importance of the topic. In Brazil, the use of annatto is so widespread that its safety is not even questioned.

Paumgartten et al. evaluated the toxicity of annatto extracts in rats. Doses up to 500 mg/kg body weight/day were introduced directly into the stomach of pregnant rats to evaluate the effect on the mother and fetus, and no adverse effects were found for either. The annatto extract did not induce an increase in the incidence of visible external, visceral, or skeletal anomalies in the fetuses. Therefore, the study suggested that the annatto extract was not toxic to rats nor was it embryotoxic. Studies performed in Brazil by Alves de Lima et al., where extracts of annatto were mixed with the food of male rats, showed that the concentrations tested had no mutagenic or antimutagenic activity in their bone marrow cells. A parallel toxicity study conducted by Hagiwara et al. showed that 0.1% annatto extract administered for thirteen weeks in the feed of male and female rats did not show any adverse effects. However, when higher doses were administered (0.3% and 0.9%), the authors noticed an increase in liver weight as well as changes in blood chemistries, including increase in alkaline phosphatase, phospholipids, and total protein, as well as albumin and albumin/globulin ratio.

Hagiwara et al. also evaluated extracts of annatto for liver carcinogenicity in rats and found no evidence of liver tumors, even when given to animals at a high dose of 200 mg/kg body weight/day, compared to an acceptable dose of 0 to 0.065 mg/kg/day, thus indicating that the danger of a hepatocarcinogenic effect in humans may be absent or negligible.

A toxicity test was performed with extracts obtained from both plant seeds and shoots, and no significant effect was observed. The experiments were performed in the United States using mice as the animal model and it was found that the LD_{50} was greater than 700 mg/kg.

https://www.hindawi.com/journals/tswj/2014/857292/

Vocabulary

pigment 英 ['pɪgmənt] 美 ['pɪgmənt] n. 颜料；色料；[生] 色素
annatto seed aril 红木籽仁
dye 英 [daɪ] 美 [daɪ] n. 染料；染色；颜色 vt. 染色；给……染色
bixin 英 ['bɪksɪn] 美 ['bɪksɪn] n. 胭脂树橙
stereochemistry 英 [stɪərɪə'kemɪstrɪ] 美 [ˌsteriːoʊ'kemɪstriː] n. 立体化学
oxidative degradation 氧化降解
carotenoid 英 [kə'rɒtɪnɔɪd] 美 [kə'rɒtn,ɔɪd] n. 类胡罗卜素
geranylgeraniol 英 [d'ʒerəniːldʒərænaɪəl] 美 [d'ʒerəniːldʒərænaɪəl] n. 牛儿基；牛儿醇
saponification 英 [səˌpɒnɪfɪ'keɪʃən] 美 [səˌpɒnəfə'keɪʃən] n. 皂化
volatile compound 挥发性化合物
aqueous 英 ['eɪkwɪəs] 美 ['ekwɪəs, 'ækwɪ-] adj. 水的；水成的
annatto 英 [ə'nætəʊ] 美 [ə'nætoʊ] n. 胭脂树；从其果实采取的黄红色染料
recrystallization 英 [rekrɪstəlaɪ'zeɪʃn] 美 [rekrɪstəlaɪ'zeɪʃn] n. 再结晶作用；重结晶；重结晶作用的
chloroform fraction 氯仿馏分
negligible 英 ['neglɪdʒəbl] 美 ['nɛglɪdʒəbəl] adj. 微不足道的；可以忽略的
antifungal activity 英 ['æntɪfʌŋgəl æk'tɪvɪtɪ] 美 [ˌæntɪ'fʌŋgəl æk'tɪvɪtɪ] 抗真菌（活）性
mutagenic activity 诱变活性；引起遗传突变的活性
cytotoxicity 英 [sɪtəʊtɒk'sɪsɪtɪ] 美 [sɪtoʊtɒk'sɪsɪtɪ] n. 细胞毒性
ethanolic 乙醇的
toxicological activity 毒理学活性
incontestable 英 [ˌɪnkən'testəbl] 美 [ˌɪnkən'testəbl] adj. 无可争辩的；不可否认的；不容置辩的
parallel toxicity study 平行毒性研究
toxicity test 毒性试验

Useful Expressions

1. substance 英 ['sʌbstəns] 美 ['sʌbstəns] n. 物质；材料；实质；内容
There's absolutely no regulation of cigarettes to make sure that they don't include poisonous substances.
根本没有法规来确保香烟不含有毒物质。
2. constitute 英 ['kɒnstɪtjuːt] 美 ['kɑːnstətuːt] vt. 构成；组成
Testing patients without their consent would constitute a professional and legal offence.
未经患者同意而对其进行检查被视为违反职业操守并触犯法律。
3. devoid 英 [dɪ'vɔɪd] 美 [dɪ'vɔɪd] adj. 缺乏；没有

I have never looked on a face that was so devoid of feeling.

我从来没有见过一张如此面无表情的脸。

4. dose 英 [dəʊs] 美 [doʊs] *n.* 剂量；药量；（药的）一服；一剂

One dose of penicillin can wipe out the infection.

一剂青霉素就能消除感染。

Questions

1. When was Bixin first isolated from the seeds of Bixa orellana?
2. How many substances have been isolated from the seeds of Bixa orellana?
3. Why do people prefer dyes of natural origin?

Chapter Five

Evaluation of Medicinal Plant Resources

5.1 Quality Evaluation of Medicinal Plant Resources

Passage 1

Species Specific DNA Sequences and Their Utilization in Identification of Viola Species and Authentication of "Banafsha" by Polymerase Chain Reaction (Extract)

Example 1

Majority of the Viola Species Specific Primers Amplify Only Corresponding Species:

PCR amplification products using a set of primers, either one of the species specific and the other consensus or both species specific primers were obtained for all five species. The PCR reaction contained 2.5 ul of 10×PCR buffer, 1 ul of dNTPs (mix of 2.5 mM each), 2—10 ng DNA, 8—10 picomole of each primer, 0.5 units of Taq polymerase and a variable concentration (0.45—1.5 mM depending upon the Viola species) of $MgCl_2$ in a 25 ul reaction mix. The DNA was denatured at 94 ℃ for 3′ and then subjected to PCR upto 40 cycles. A last extension cycle of 2 minutes in each case was also given at 72 ℃. The annealing and extension temperatures and the times along with $MgCl_2$ concentration were optimized for each of 5 Viola species with each set of primers and are discussed below.

Viola odorata: Using 8 picomole each of the forward and reverse primers (BT61.F and BT61.R) respectively along with annealing at 68 ℃ for 15 seconds, extension time of 18 seconds and a concentration of 0.85 mM of $MgCl_2$ was optimized to give a band of 150 bps as expected with only V. odorata but not with any of the other plants. There was also a similar amplification product when the genomic DNA of all the five plant species were mixed in equal ratio. The PCR product was V. odorata genome specific and could be detected reproducibly in mixture of all the 5 plant species. There was no significant difference in the band intensity when the genomic DNA from fresh tissues or the dried tissues were taken. Increasing $MgCl_2$ concentration or the annealing time led to the appearance of a few faint bands of high molecular weight.

Viola canescens: Use of BT71.F and BT71.R as forward and reverse primers respectively along with an annealing temperature of 62 ℃ for 24 seconds and an extension time of 17 seconds along with a $MgCl_2$ concentration of 1.5 mM gave an expected band of

200 bps with only V. canescens as well as in the mixed samples of all 5 Viola species but not individually with any of the other 4 DNA templates. Increasing annealing temperature beyond 62 ℃ led to no amplification while reducing temperature led to appearance of several faint bands in addition to 1 prominent expected band.

Viola pilosa: Similarly use of BT811.F and BT811.R set of primers along with cycling parameters of annealing at 62℃ for 22 seconds and extension for 17 seconds along with a $MgCl_2$ concentration of 1.1 mM gave a band of expected size of 200 bps with V. pilosa and V. canescens but not with any of the other plants tested. Similar amplification product was obtained when the genomic DNA of all the five plant species were mixed in equal ratio (lane X). To demonstrate that V. pilosa specific primers detect both V. pilosa and V. canescens but V. canescens specific primers detect only V. canescens specifically, the applicants used species specific primers for these species to amplify mixtures of 4 Viola species DNA templates minus V. pilosa or V. canescens. Pilosa specific primers detect both V. pilosa and V. canescens but not vice versa.

Viola betonicifolia: Using BT91.F and M28 as forward and reverse primers respectively along with an annealing temperature of 52 ℃ for 30 seconds, an extension time of 17 seconds, and $MgCl_2$ concentration of 0.55 mM gave an expected band of 311 bps with only V. betonicifolia or in a mixture of all 5 species but not with any of the other plants individually.

Viola tricolor: Using BT101.F and M28 set of primers along with an annealing temperature of 52℃ for 30 seconds, an extension time of 17 seconds, along with a $MgCl_2$ concentration of 0.45 mM gave an expected band of 190 bps with only V. tricolor or in a mixture of all 5 species but not with any of the other plants. Surprisingly V. tricolor and V. betonicifolia required almost similar amplification conditions.

Example 2

Conserved 5S rRNA Gene Based Primers (3′ and 5′) Gave an Identical Amplification Product in a Population of 28 Individual Plants of Viola pilosa:

In order to find out 5S rRNA associated spacer length and sequence variability within the individuals of a species, the applicants analyzed 28 individuals plants of V. pilosa selected randomly from 2 plots (from IHBT campus, Palampur and Jogindernagar Herbal Garden, Jogindernagar), for their amplification products using consensus 5S rRNA primers M27 and M28. All the plants gave identical banding patterns. When these plants were subjected to RAPD analysis using OPE-09 primer they again showed identical banding patterns. When 2 more primers were analyzed they showed some differences in many of the plants as expected.

Example 3

Viola betonicifolia Specific Primers Gave an Identical Amplification Product in 12 Individual Plants Collected from Nature:

Similar to Example 1, 12 individual plants of V. betonicifolia were analyzed for their amplification products using V. betonicifolia species specific set of primers. All the plants gave identical banding patterns. These PCR products when denatured and analyzed in a 1.8% agarose gel again revealed identical banding patters. When these plants were subjected to RAPD analysis using OPE-09 primers they showed little differences, for example, in lane No. 2 and 3 the 3rd band from top is present only in 4 samples which is different in size in lane No. 8.

Example 4

Market Samples of Banafsha Revealed that Majority of Them Were Fake Samples:

In order to find the genuineness of banafsha, the applicants analyzed 18 market samples of banafsha collected from different markets during 1997—1999 for their amplification products using V. canescens specific primers. Some evidence shows that 7 samples were positive for V. canescens which does not represent genuine banafsha. This example clearly demonstrates that the developed PCR band approach works well for market samples.

The main advantages of the present invention are the following:

(i) It is specific to Viola odorata, V. pilosa, V. canescens, V. betonicifolia and V. tricolor.

(ii) It is highly sensitive and only nanogram amounts of DNA is required.

(iii) It can work equally well for degraded DNA.

(iv) Only mg amounts of samples are required.

(v) It can work well for the processed and powdered samples.

(vi) It can detect presence of Viola species even in admixtures of samples and herbal formulations.

(vii) The presence of Viola species specific PCR products can be visualized in a simple agarose gel and no hazardous radioactive labeling or time consuming and complex systems are needed.

(viii) It is rapid.

(ix) It has a potential for automation.

http://xueshu.baidu.com/s? wd=paperuri%3A%28b4ca5faaee03e932bd2c138a0b0 2e77e%29&filter=sc_long_sign&tn=SE_xueshusource_2kduw22v&sc_vurl=http%3A %2F%2Fwww.freepatentsonline.com%2F6924127.html&ie=utf-8&sc_us=155057936 29619604298

Vocabulary

consensus 英 [kən'sensəs] 美 [kən'sɛnsəs] n. 一致；舆论；一致同意；合意
reproducibly 英 [ˌriːprə'djuːsəbli] 美 [ˌriːprə'djuːsəbli] adv. 可再生产地
intensity 英 [ɪn'tensəti] 美 [ɪn'tɛnsɪti] n. 强度；烈度；强烈

prominent 英 [ˈprɒmɪnənt] 美 [ˈprɑːmɪnənt] adj. 著名的；突出的；杰出的；突起的
degraded DNA   降解 DNA
formulation 英 [ˌfɔːmjʊˈleɪʃn] 美 [ˌfɔrmjəˈleʃən] n. 配方；构想；规划；公式化
hazardous radioactive labeling   危险放射性标签

Useful Expressions

1. detect 英 [dɪˈtekt] 美 [dɪˈtɛkt] vt. 查明；发现；洞察
Arnold could detect a certain sadness in the old man's face.
阿诺德能觉察到老人脸上的一丝悲伤。
2. be subjected to   受；遭受
Mental patients should be subjected to strict surveillance.
对精神病人应严加看管。
3. vice versa   反之亦然；反过来也一样
Teachers qualified to teach in England are not accepted in Scotland and vice versa.
在英格兰有教书资格的老师在苏格兰得不到认可，反之亦然。

Questions

1. How many species of viola were used in the examples? What are they?
2. What does the experiment on market samples of banafsha show?
3. What are the main advantages of the invention?

Passage 2

Quality Evaluation of Ayurvedic Crude Drug Daruharidra, Its Allied Species, and Commercial Samples from Herbal Drug Markets of India (Extract)

Introduction

Berberis aristata known as "Daruharidra" in Ayurveda is a versatile medicinal plant used singly or in combination with other medicinal plants for treating a variety of ailments like jaundice, enlargement of spleen, leprosy, rheumatism, fever, morning/evening sickness, and snakebite, and so forth. In addition, the decoction of root or stem of Berberis known as "Rasaut" is specifically used in eye disease, skin disorders, and indolent ulcers. Its use in the management of infected wounds has also been described in Ayurvedic classical texts. The major alkaloid of the plant is berberine, which is known for its activity against cholera, acute diarrhea, amoebiasis, and latent malaria and for the treatment of oriental sore caused by Leishmania tropica.

Although the roots of B. aristata are considered as the official drug, the study revealed that different species of Berberis, namely, B. asiatica, B. chitria and B. lycium are also used as Daruharidra in different parts of the country. In southern India, however, Coscinium fenestratum is used as "Daruharidra". The study also shown that most of the market material sold as Daruharidra consists of mostly the stem parts than the roots of Berberis species.

As such there are different alkaloids available to differentiate different Berberis species. Several workers have also done molecular analysis of different Berberis species including the presented four species which reflects the use of molecular markers and sequence analysis for identification at inter-and intra-specific level.

Over exploitation of B. aristata created scarcity of the material that opened new vistas to identify a possible substitute for this species. During the market surveillance of different herbal drug markets of India, it was observed that almost all the markets either comprise Berberis lycium or Berberis asiatica. Although a detailed pharmacognostic study of B. aristata, B. asiatica, and B. chitria is reported by Srivastava et al., market surveillance is not yet performed. Hence, the present study has been undertaken, which may be useful to pharmaceutical industries for the authentication of the commercial samples and to explore the possibilities of using other species as a substitute of B. aristata.

Materials and Methods

The plant materials were collected from the Dhanaulti (Uttaranchal) region of India and the roots were preserved in 70% ethyl alcohol for histological studies. Procurement of commercial samples was done from various important drug markets of India, namely, Aligarh, Amritsar, Bangalore-I, Bangalore-II, Delhi, Hyderabad, Jammu, Lucknow, Trichur, Varanasi, and so forth.

Microtome sections were cut and stained with safranin and fast green and photographed with Nikon F70X camera. Physicochemical and phytochemical studies like total ash, acid insoluble ash, tannins, and total alkaloids were calculated from the shade dried powdered material according to the recommended procedures. Further, heavy metal studies and quantitative estimation of berberine through HPTLC have also been performed as per ICH guidelines.

Results and Discussion

Morphological studies showed certain minor variations in all the four Berberis species. For example, in B. aristata and B. chitria the cut surface is bright yellow while that of B. asiatica and B. lycium is lemon yellow, and deep yellow, respectively. Similarly, the colour of wood bark has also minor variation, namely, it is yellowish brown to yellowish gray in all the three species except in B. lycium the colour is grayish white. Likewise the numbers of pericyclic fibres are different in all the four species, for example, the maximum is found in B. aristata and minimum in B. lycium.

A comparative account of all physicochemical values has been depicted in histograms. It is quite clear from these studies that no significant variation was observed in total ash of all the four species of Berberis. However, the percentage of acid insoluble ash of roots and stem showed significant variations; for example, the highest percentage of 0.26% acid insoluble ash was observed in B. asiatica root and the lowest one of 0.05% was noted in B. aristata root. It is interesting to note that the percentage of alcohol and water-soluble extractives were higher in root as compared to stem except in B. chitria. On the contrary, the percentage of starch was higher in stem (14%—19%) except in B. lycium (root) and it was 26.03%. Percentage of tannin was more or less similar in both in root and stem of all the samples (0.7%—1.7%).

The percentage of total crude alkaloid percentage was also estimated and it was found that it varied from species to species, that is, maximum in roots of B. chitria (3.65%) followed by the roots of B. lycium (2.8%), B. aristata (2.45%), and B. asiatica (2.4%), respectively. Besides, the active constituents berberine one of the major alkaloids was also calculated through HPTLC densitometric method (solvent system, npropanol : water : formic acid, 90 : 80 : 0.4) and it was found more in roots as compared to stem, that is, 2.25%—5.20% and 1.02%—2.01%, respectively. Its concentration was also varied from species to species, that is, maximum in roots of B. chitria (5.20%) followed by B. lycium (3.99%), B. aristata (3.55%), and B. asiatica (2.25%).

A comparative study of official drug B. aristata sample with that of commercial samples was made and it was found that the Bangalore-I sample has all the similar morphological characters of roots of B. asiatica, namely, (i) outer surface grayish brown with 2 mm thick friable bark, which was separated out immediately leaving muddy yellow surface of the wood; (ii) transversely cut surface lemon yellow; (iii) sclerieds mostly in groups of 2—12 rarely solitary and comparatively more than other three species; (iv) pericycle fibres interrupted by stone cells; (v) length of the vessel elements much more than the other species, that is, up to 500 um (±181.0); (vi) physicochemical values are within the prescribed range of Ayurvedic Pharmacopoeia of India, hence, identified as roots of B. asiatica.

Similarly, majority of anatomical characters of Aligarh and Varanasi samples matched with the stem and roots of B. asiatica in having (i) some pieces with fine longitudinal ridges and small warts on the outer surface of bark and dark brown outer surface of wood; (ii) transversely cut surface lemon yellow; (iii) sclerieds rarely solitary mostly in groups of 2—12 and comparatively more than other three species; (iv) pericycle fibres interrupted by stone cells; (v) length of the vessel elements much more than the other species up to 600 um (±181.0); (vi) trachieds up to 680 um (±167.0) long; (vii) some other pieces have grayish brown outer surface with 2 mm thick friable bark which was separated out immediately leaving muddy yellow surface of the wood; presence of prominent pith as in stem of B. asiatica.

Furthermore, the commercial samples of Delhi and Lucknow showed close resemblance with the stem of B. asiatica by the presence of (i) fine longitudinal ridges and small warts on the outer surface of bark and yellowish creamy transverse cut surface; (ii) dark brown outer surface of wood as appeared after peeling off the bark; (iii) sclerieds rarely solitary mostly in groups of 2—12 and comparatively more than other three species; (iv) pericycle fibres interrupted by stone cells; (v) trachieds up to 680 μ (± 167.0) and vessels up to 600 μ (± 102.0) long (vi) pith.

Similarly, the market samples of Amritsar and Jammu were found to be the mixture of stem and root of two different Berberis species. Amritsar samples were found to be the stem of B. aristata and root of B. asiatica while Jammu sample comprised of root of B. chitra and stem of B. asiatica.

The morphological characters in Amritsar sample are (i) outer surface creamish brown with knots, fine longitudinal ridges, and flakes; (ii) bark very thin and brittle; (iii) transverse cut surface bright yellow; (iv) outer surface of wood which appeared after peeling off the bark was yellowish brown; (v) sclerieds solitary or in a group of 2—10; (vi) pericyclic fiber mostly solitary, rarely in groups of 2—10; (vii) length of the fibres much more, that is, about 630μm as compared to other three species.

https://www.ncbi.nlm.nih.gov/pmc/articles/PMC3566491/

Vocabulary

berberis 英 [bɜːbəris] 美 [bɜːbəris] *n.* 小檗属植物
versatile 英 ['vɜːsətaɪl] 美 ['vɜːrsətl] *adj.* 多才多艺的；多功能的
jaundice 英 ['dʒɔːndɪs] 美 ['dʒɔndɪs, 'dʒɑn-] *n.* 黄疸病；偏见
spleen 英 [spliːn] 美 [splin] *n.* 脾
leprosy 英 ['leprəsi] 美 ['lɛprəsi] *n.* 麻风病；大麻风
rheumatism 英 ['ruːmətɪzəm] 美 ['rumətɪzəm] *n.* [医]风湿病
decoction 英 [dɪ'kɒkʃən] 美 [dɪ'kɒkʃən] *n.* 煎煮；煮出的汁；煎熬的药
alkaloid 英 ['ælkəlɔɪd] 美 ['ælkəlɔɪd] *n.* 生物碱；植物碱基
berberine 英 ['bɜːbərɪn] 美 ['bɜːbərɪn] *n.* 黄连素
cholera 英 ['kɒlərə] 美 ['kɑːlərə] *n.* [医] 霍乱；虎疫
acute diarrhea 急性腹泻
amoebiasis 英 [əˈmiːbɪəsɪs] 美 [əˈmiːbɪrsɪs] *n.* 阿米巴病；变形虫病
surveillance 英 [sɜːˈveɪləns] 美 [sɜːrˈveɪləns] *n.* 盯梢；监督
safranin 英 ['sæfrənɪn] 美 ['sæfrənɪn] *n.* 盐基性红色染料；番红精
pericyclic 英 [ˌperɪ'saɪklɪk] 美 [ˌperɪ'saɪklɪk] *n.* [植]中柱鞘
physicochemical 英[ˌfɪzɪkəʊ'kemɪkəl] 美 [ˌfɪzɪkoʊ'kemɪkəl] *adj.* 物理化学的
histogram 英 ['hɪstəɡræm] 美 ['hɪstəɡræm] *n.* 柱状图
transversely 英 [trænsˈvɜːsli] 美 [trænsˈvɜːrsli] *adv.* 横着；横切地；横断地
pith 英 [pɪθ] 美 [pɪθ] *n.* 精力；(木)髓

creamy 英［ˈkriːmi］美［ˈkrimi］adj. 奶油色的；多乳脂的或似乳脂的
longitudinal ridges 纵脊；纵行脊
flake 英［fleɪk］美［flek］n. 小薄片
brittle 英［ˈbrɪtl］美［ˈbrɪtl］adj. 易碎的

Useful Expressions

1. scarcity 英［ˈskeəsəti］美［ˈskersəti］n. 稀少；不足；缺乏；萧条
This scarcity is inevitable in less developed countries.
这一匮乏在发展中国家不可避免。
2. depict 英［dɪˈpɪkt］美［dɪˈpɪkt］vt. 描述；描绘；描画
Margaret Atwood's novel depicts a gloomy, futuristic America.
玛格丽特·阿特伍德的小说描述了一个黑暗无望的未来美国。
3. transverse 英［ˈtrænzvɜːs］美［ˈtrænzvɜːrs］adj. 横向的；横断的；横切的
The distribution was the most in the ascending colon, transverse colon, descending colon.
分布以升结肠、横结肠、降结肠多见。

Questions

1. What illness can be treated with berberis?
2. What species of berberis can be used as medicine?
3. What does the comparative study of all physicochemical values show?

5.2 Benefit Evaluation of Medicinal Plant Resources

Passage 1

Sustainable Utilization of traditional Chinese medicine Resources: Systematic Evaluation on Different Production Modes (Part 1)

Introduction

traditional Chinese medicine (TCM) recently is widely accepted by patients and attracting more and more attention of researchers with the change of disease modes and the rise of "return to nature" in the world. Currently 45% of all countries (regions) in the world are using TCM, and the global trade of TCM has reached 40 billion USD a year, with an increasing rate of 10% per year. China exported 25% of the global demand of TCM while 75% of which was raw materials. Currently, a total of 80% of TCM are from continuous wild collection without scientific plans. The natural reserves can hardly meet the rapidly increasing demand. At the same time, the wild herbal resources are quickly decreasing by 30% every year. Consequently 80% of the most usually used species cannot meet medical demand. Data analysis showed that 1,800—2,100 medicinal species were facing the challenge of extinction in China. Even though some of the wild herbs can recover naturally within 3—5 years, the recovery speed falls much behind the one of the rising demand. As a result, TCM resources are facing more and more challenges of sustainable utilization.

Long-term sustainable utilization of TCM resources should combine market demand of raw materials, ecological stability, and social benefits. Currently cultivation and natural fostering were the main production modes to ease contradictions between the decrease of natural reserves and the increase of market demand of wild medicinal resources. Cultivation can be implemented in a large scale and is an efficient method to rapidly provide sufficient urgent raw medicinal materials compared with other methods. However, cultivation requires massive cultivated land and plant disease and pests, heavy metal and pesticide residue are the major obstacles to limit its application to all medicinal plants. Moreover, the quality of its output (raw materials) sometimes cannot satisfy the clinical criteria. Natural fostering, also named as wild nursery or semi-imitational cultivation, is a new kind of herb production method. It can combine economic benefit and diversity protecting practically and can solve the problem between subsistence and biodiversity conservation effectively. In China emphasis is given to natural fostering which aims to maintain and recover viable populations of wild species in nature. But it still has limitations such as the long-time process and low output, which cannot meet the rapid increase of market demand in short term. Though wild collection, natural fostering, and cultivation have own advantages in yielding raw medicinal materials, they cannot completely solve the current problem of herbal resource sustainable utilization alone. How to choose production mode depends on natural reserves, usage amount, and biological characteristics of medicinal plant species. This study performs a systematic evaluation on these three production modes in different dimensions including current application status, technological challenges, input-output ratio, and ecological impact. In addition, we present illustrations to indicate the characteristics of each method, which can practically guide the selection of TCM production modes for resource sustainable utilization.

Current Status of Different Production Modes

1) Wild Collection

TCM resources mean the healing herbal materials. According to the statistics, there are 11,146 medicinal plants species, belonging to 383 families, and 2,313 genera. The herbal geographical distribution covers different longitudes, latitudes, and altitudes in China. Different ecological habitat causes different genuine medicinal materials. Nowadays there were 100,000 TCM prescriptions and these prescriptions used 700 Chinese herbal species, 80% of which come from wild collection such as Polygonum cuspidatum,

Leonurus japonicus, Forsythia suspensa, and Bupleurum chinense. The mostly used Chinese herbs need wild collection. Due to the finite herbal storage, the increasing demand, and the harsh living conditions, the output of wild herb collection is reducing every year.

Even though the Chinese government has started to improve ecological environment to protect the Chinese herbal habitat, the increasing demand of the whole world market still makes a great deal of medicinal plants face the possible extinction. The problems including the lack of wild herbal collecting plan, biomes' destruction, and degraded ecologic environment are becoming more and more serious. Some famous wild herbs, such as Cordyceps sinensis, Dendrobium officinale, Fritillaria cirrhosa, and Saussurea involucrata, are becoming more and more difficult to be found in wild habitat. Although some medicinal plants have been successfully cultivated, their wild species can still no longer be found within the latest decades of years, just like ginseng, notoginseng, and Gastrodiae. It would lead to great obstacles in future when these genuine medicinal plants need to be selected from wild resources for breeding.

The main reasons that endanger the wild Chinese herbal resources are the following. Firstly, the national and international market demand is boosting. There are about 1.368 billion people in China in 2014. This is a huge consumption group. Besides, Chinese herbal trade has extended to 120 countries in the world market. The herbal varieties have reached 500 species and mainly are transported abroad as raw materials. Secondly, the worsening global climate and Chinese ecological environment are threatening medicinal plant habitat. In the past 30 years the rapid industrialization of China has caused a huge pressure to wild environment. The changing environment has lowered the recovery speed of wild herbal plants. Some herbal plants even cannot be recovered at all because of the damage of their natural environment. Thirdly, wild collection was not scientifically carried out. In China the people who work on wild herb collection have different education levels. Some of them have little knowledge on herbal sustainable utilization. They tend to follow their own habits and economic motivations to collect wild herbs, which causes that some herbs and their habitats are destroyed destructively. For example, digging a wild licorice has to destroy accompanying plants of 10 m^2. And digging up wild plants of Ephedra species destroys 3,200 hm^2 meadow every year. Therefore, current wild collection cannot guarantee the sustainable development of Chinese herbal resources.

2) Natural Fostering

Natural fostering mainly focuses on increasing the number of herbal population to provide raw medicinal materials. This method should be implemented in original habitats, which is different from artificial cultivation. Natural fostering can effectively combine medicinal plant production and economic benefit. It can increase the recovery ratio of original population. Finally, it does not change the basic community trait of original habitat. Therefore, natural fostering unites industrialized production of TCM and ecological protection. Natural fostering has been carried out in many Chinese herbs, such as Fritillaria cirrhosa, F. unibracteata, Glycyrrhiza uralensis, G. inflate, G. glabra, Panax ginseng, Ephedra sinica, E. intermedia, Coptis chinensis, C. deltoidea, C. teeta, Gastrodia elata, Saussurea involucrata, and Cordyceps sinensis. Practice proves it to be a pragmatic way to produce TCM materials and conserve biological diversity.

Natural fostering is mostly suitable for such herbs. Firstly, their original habitats are special and cultivation cost is very high. Secondly, their commercial characters and quality have great variations after being cultivated. Thirdly, their wild distribution areas are concentrated and great achievement of production will be made by natural fostering.

Natural fostering is an innovative method for Chinese herbal production. More than 19 Chinese herbs have been used to produce raw materials through natural fostering and among which 12 herbs have realized large-scale production. The key advantage of natural fostering is that the herbal quality from its output is very close to that from wild collection. For example, the polysaccharide content of wild Ranunculus ternatus is 14.1% and 10% from natural fostering. However, the content of total amino acids of Ranunculus ternatus is 2%—3% higher from natural fostering than from wild collection. But they almost have the same kinds of amino acids. Similarly, wild Cordyceps sinensis have the same varieties of amino acids as that from natural fostering and both of their adenosine contents are more than 0.1 mg/g, which meets the quality requirement of Chinese Pharmacopoeia. The profit motivation is the main drive to prompt peasants to implement natural fostering of medicinal plants. For example, the income from fostering Coptis is 15 times higher than the one from planting crops.

Natural fostering is a promising herbal producing method which is a combination of wild collection and cultivation. But its technology is still not mature and the germ plasm for natural fostering has not been identified completely. The production scale is not as large as artificial cultivation and stays at the primary phrase. Although there are numbers of experimental projects of natural fostering in China, they are limited by the weakness of basic studies and long-term process. The success of natural fostering also depends on the further study of specie characteristics.

3) Cultivation

Herbal cultivation is one of the most effective methods which can not only satisfy market demand but also release the ecological pressure caused by wild collection. In China, the area of herbal cultivation has increased from 400,000 hm^2 in 1950s to 9,330,000 hm^2 in 2015. There are altogether 200 herbs that can be artificially cultivated, 100 of which have achieved large-scale cultivation including Eucommia ulmoides, Magnolia officinalis, Bupleurum chinense, and Platycodon grandiflorum. The output of herbal production by cultivation has reached 400,000 tons per year. The yield of 200 herbs usually uses Chinese herbal medicine from cultivation accounts for more than 60% of the whole market demand per year in China. In particular some herbs such as ginseng and notoginseng are provided absolutely by cultivation. More and more companies are beginning to recognize the supply crisis of raw medicinal materials, and the Good Agriculture Practice (GAP) is implemented widely.

TCM cultivation not only provides raw medicinal materials but also brings additional essential problems such as excessive heavy metal and pesticide residues. Although the number of herbs which can be cultivated is increasing, artificial breeding has been carried out on only 20 kinds of herbs. The degeneration of germ plasm leads to plant diseases and insect pests and significant output reduction. The planting area in 2008 decreased by 40% compared with the area in 2002. Plant diseases and insect pests are two of the main reasons. Another problem of herbal planting is the lack of scientific design due to profit issues. When planting fruits, crops, and economic forest can bring more income than planting herbs, farmers will give up herbal cultivation. For example, the scale of Polygonum multiflorum in 1977 had reached 453.3 hm^2 in Deqing County in Guangdong Province while the planting area decreased by 90% in 2012 because planting citrus and other fruits can bring more income.

Although herbal cultivation can increase production and economic profit rapidly in single population, it is not a sustainable way for the development of herbal resources. Compared with crop cultivation which aims to get the first metabolite (protein, fat, sugar, etc.) of the plants, the main purpose of herbal production is to produce the secondary metabolites (such as alkaloids, saponins, terpenes). Herbal cultivation goes against plant natural growing regularity by escaping from community environment. It is suggested that improving cropping ratio and intercropping ratio may be able to solve the current problems in herbal cultivation.

https://www.hindawi.com/journals/ecam/2015/218901/

Vocabulary

reserve 英 [rɪˈzɜːv] 美 [rɪˈzɜːrv] n. 储备；保留；保护区
massive 英 [ˈmæsɪv] 美 [ˈmæsɪv] adj. 大规模的；大的；重的
pesticide 英 [ˈpestɪsaɪd] 美 [ˈpestɪˌsaɪd] n. 杀虫剂；农药
viable 英 [ˈvaɪəbl] 美 [ˈvaɪəbəl] adj. 切实可行的；能自行生产发育的
notoginseng 英 [nəʊˈtɒdʒɪnseŋ] 美 [noʊˈtɒdʒɪnseŋ] n. 三七
natural fostering　自然养育
amino acid　氨基酸
adenosine 英 [əˈdenəsiːn] 美 [əˈdenəsɪn] n. 腺苷
citrus 英 [ˈsɪtrəs] 美 [ˈsɪtrəs] n. [植] 柑橘属果树；柠檬；柑橘
intercropping 英 [ɪntəˈkrɒpɪŋ] 美 [ɪntəˈkrɒpɪŋ] n. 间混作

Chapter Five Evaluation of Medicinal Plant Resources

Useful Expressions

1. implement 英 ['ɪmplɪment] 美 ['ɪmpləmənt] *vt.* 实施；执行；使生效；实现
The government promised to implement a new system to control financial loan institutions.
政府许诺实施新的制度来控制金融贷款机构。
2. residue 英 ['rezɪdjuː] 美 ['rezɪduː] *n.* 残余；残渣；余渣；残余物
Always using the same shampoo means that a residue can build up on the hair.
总是用同一种洗发水就意味着某种残留物会在头发上越积越多。

Questions

1. Why are TCM resources facing more and more challenges of sustainable utilization?
2. What should be done to keep long-term sustainable utilization of TCM resources?
3. Please describe briefly the current status of TCM production modes.
4. What the main reasons that endanger the wild Chinese herbal resources?

Passage 2

Sustainable Utilization of traditional Chinese medicine Resources: Systematic Evaluation on Different Production Modes (Part 2)

Technological Challenges

1) Wild Collection

The key technological points of wild collection include collection methods, transportation, species identification, and collecting period. Among these factors, one of the most crucial technological difficulties is collection method. We should carefully design collection method to avoid possible damage on surrounding plants. At the same time, another difficulty of wild collection is transportation. The original distribution of wild medicinal plants is usually located in remote mountainous areas with poor establishments whose inconvenient traffic and poor information systems make it challenged to transport the wild collected herbs. For example, wild Liquorice, Ephedra, and Cistanche mainly distribute in desert areas and are difficult to be transported. Moreover, long transportation process may also have influence on the quality of Chinese materia medica.

2) Cultivation

Germ plasm selection, breeding, fertilization, and prevention of diseases and insect pests should be paid more attention to in herbal cultivation.

(1) Fertilization

Currently few basic studies on agrology of TCM are systematically carried out. We still do not know much about suitable soil conditions, balanced fertilization technology, and the relationship between soil environment and herbal intrinsic quality. Fortunately, the varied soil conditions are being gradually recognized. Researchers found that, in the 29 cultivation areas of Paris polyphylla Smith var. yunnanensis in nine cities of Yunnan Province, the nutritional soil status was extremely uneven. Soil nitrogen in eight of these cultivated areas was below the normal level, which would affect herbal growth and output in the next year. The absorption of phosphorus and potassium was also influenced. The pH value of four cultivation areas was beyond the optimal growth range (4.5 to 6.3). People have gradually realized that the rhizome of some herbs has the ability to enrich heavy metal elements. Even though the content of heavy metals is in a low level in soil, the actual amount of heavy metal in herbal materials still exceeds the standard. Han et al. found that the average content of copper, lead, arsenic, cadmium, and mercury exceeded the standard 21.0%, 12.0%, 9.7%, 28.5%, and 6.9%, respectively, after analyzing 312 kinds of Chinese herbs.

(2) Pest Prevention

The basic researches on pest prevention of Chinese herbal species remain limited until recently. The studies on the relationship among soil herbs and germ, as well as the physiological, biochemical, and molecular mechanic researches, are also in infancy. Because there are many differences between Chinese herbal species and crops including growth habit, stress resistance, and main target products, Chinese herbal cultivation cannot just apply the techniques used in ordinary crops completely. According to an investigation analyzing 300 kinds of Chinese herbal materia medica, the majority of the samples have residue of organic chlorine pesticides which can lead to hepatomegaly, degeneration of liver cell, damage of central nervous system, and bone marrow.

(3) Seed Selecting and Breeding

Improper selection of original herbal species will lead to species confusion and reduction of diversity. For example, seeds from all kinds of Cannabis can be used as Huomaren materia medica, which causes the confusion of Huomaren germ plasm. Due to the divergence of stress resistance and growth habit, it is greatly difficult to introduce and cultivate Cannabis sativa according to the same protocol. The most difficulty to select medicinal varieties is the intraspecies variation. Different classifications are divided according to the variation including subspecies, variety, and variant, which causes the quality divergence of herbal medicine and different clinical efficacy. It is difficult to find good variety possessing not only the highest yield but also the best quality. Unfortunately there have been no new varieties with stable genetic property up to now.

3) Technological Challenges for Natural Fostering

The key technological points of natural fostering include selection of site, construction of seedlings base, density adjustment, and scientific harvesting. Among these technological points, the main challenges for natural fostering include the following.

(1) Selection of Site

Natural fostering requires that the site should be in the original habitat or the ecological conditions similar to the original habitat areas. To determine whether an area is suitable for breeding, the most reliable and effective way is to carry out a longer period of experiment. However, it needs a lot of manpower, material, and observation for up to several years of growth cycle, and in practice the experiment is much difficult to carry out in a large scale. A wild habitat may not be suitable for the growth of target medicinal species; at least most of the plants cannot be guaranteed to live in optimal living conditions and toward the development of population growth. So it is difficult to develop natural fostering to reach a large industrial scale in some situations. Multicriteria assessment is essential to select suitable fostering site: (i) direct information on species distributions; (ii) market analysis on potential medical plants; (iii) community types or biotic units according to the evaluation of effective components; (iv) transportation convenience; and (v) other goals.

(2) Construction of Seedlings Base

The key part of base construction is how to produce vigorous seedlings or healthy seeds. Original funds and technologies are very important. The aims of nursing seedlings are somewhat different from breeding in agriculture. Two methods are usually adopted: selection of germ plasm resources and crossbreeding. Selections of germ plasm resources are from wild species whose characters include ability of resisting adversity, high production, high content of effective chemical component, and so on. Sometimes, the selection needs to be adjusted in terms of relative medical aims such as medicinal organs (flower, radix, rhizome, and leaf), characters of effective component (volatile, poisonous), and values of medical goods (genuine, shape). Crossbreeding can be operated according to common approaches in agriculture. One point needs to be announced that medical raw materials from crossbreeding have not been verified through long-term experiments and not yet accepted by traditional Chinese herbalist doctors. From this view, crossbreeding in natural fostering is used in yielding of medical materials for component of extraction, not for "yinpian" (semimanufactured goods for medicine through different physical methods). In addition, building and field construction are absolutely necessary.

(3) Scientific Harvesting

Three parameters should be taken into account in natural fostering: quantity to harvest, time of harvest, and the condition of the plant community. The quantity of picking should not affect community structure and not pick in excess to ensure sustainable utilization in following years. It is absolutely necessary to pick medicinal materials in proper time because the content of effective chemistry component is different in different stages. As the extension of fostering time, herbal population density, medicinal ingredients, and biomass gradually increase. However, natural fostering may cause the degradation of competitiveness in licorice population after 5 years. For some herbs, the shorter the growing years, the higher the content of the active ingredient, such as calycosin glucoside and formononetin in Astragalus membranaceus. Although early acquisition of these herbs could get better economic benefits, it may bring ecological loss. On the contrary, for some herbs the longer the growing years, the higher the content of medicinal ingredient, such as ginseng. In order to achieve bigger production and high content of component, studies should be carried out on content dynamic curves of aimed medicinal materials in its life. The final objective is to find out key point of intersection between quantity and content to get the biggest effective biomass. All of the operations should preserve ecological stability, after all, which is our objective.

Economic Input and Output

1) Wild Collection

Wild collection depends on directly picking raw medicinal materials from natural resources. Therefore, manpower is the main economic input. Local people do not need to invest any other economic resources to conduct wild collection, which generates great motivation for wild collection without considering systematic plan and the damage of ecological environment. Li et al. found that the annual income of farmers in Zhouzhi County was 1,767.7 RMB/household in 2007 only by wild herbal collection. Another study found that the annual income was 1,200 RMB/person in Ussuri area of Heilongjiang Province, which stimulated more than 3,000 persons to join in wild herbal collection.

2) Cultivation

The economic input of herbal cultivation includes land leasing, buying seeds, irrigation, fertilizers, pesticides, and manpower. Generally, the bigger the planting area, the higher the yield, the higher the economic income, and the higher the land cost. Most of leasing lands are located in the main producing areas, and the rents are different among different regions. According to the statistics of the National Agricultural Cost-benefit Data Assembly, the land rents of ginseng in Jilin, Liaoning, and Heilongjiang Province were 9.07 RMB/m^2, 0.54 RMB/m^2 and 0.46 RMB/m^2, respectively. Moreover, the land rents of cultivating different herbs in the same province were also different. For example, in Hubei Province the average land rent of Coptis was 0.14 RMB/m^2 but 0.07 RMB/m^2 for Kikyo. In addition, different planting methods also result in different input-output ratios. Evidence shows that the input-outcome differences between the standard planting (demonstration group) and nonstandard planting (control group) for Pinellia are obviously apparent.

Comparing the two planting methods, there are significant differences between the cost of seeds, organic fertilizer, and employee. And the input-output ratio of standard planting methods is higher than the one of nonstandard planting. It indicates that more attention should be paid to seeds, soil management, and fertilization in herbal cultivation.

3) Natural Fostering

The economic input of natural fostering includes buying seeds, base construction, and harvest. Different fostering modes can bring different economic benefits. New evidende shows that planting Coptis using three models (cultivation in greenhouse, fostering under Cryptomeria japonica and Coriaria nepalensis forest, resp.) causes different input-output ratios. Cultivation in greenhouse requires highest cost and fostering under Cryptomeria japonica forest brings the highest yielding. Natural fostering has a great advantage in planting Coptis. It is proved that natural fostering has higher benefit than cultivation in ginseng production. Compared with herbal cultivation, natural fostering is mainly implemented in mountainous areas, which causes higher cost of employees due to the poor traffic. As the examples of City and Rhizoma Paridis fostering in Dujiangyan City of Sichuan Province, the human cost of transportation is 170 RMB from picking point to collection area. Therefore, Chinese medicine enterprises have to build their factories close to the fostering area to reduce transportation cost.

https://www.hindawi.com/journals/ecam/2015/218901/

Vocabulary

materia medica　本草
fertilization 美 [ˌfɚtlɪˈzeʃən] *n.* 施肥；受精
intrinsic 英 [ɪnˈtrɪnsɪk] 美 [ɪnˈtrɪnzɪk, -sɪk] *adj.* 固有的；内在的；本质的
phosphorus 英 [ˈfɒsfərəs] 美 [ˈfɑːsfərəs] *n.* [化] 磷；磷光体
potassium 英 [pəˈtæsiəm] 美 [pəˈtæsiəm] *n.* [化] 钾
physiological 英 [ˌfɪziəˈlɒdʒɪkl] 美 [ˌfɪziəˈlɑdʒɪkəl] *adj.* 生理的；生理学的
bone marrow　骨髓
divergence 英 [daɪˈvɜːdʒəns] 美 [daɪˈvɚːdʒəns] *n.* 分歧；背离；分叉；离题
vigorous 英 [ˈvɪɡərəs] 美 [ˈvɪɡərəs] *adj.* 有力的；精力充沛的
seedling 英 [ˈsiːdlɪŋ] 美 [ˈsidlɪŋ] *n.* 实生苗；秧苗；幼苗；树苗
ingredient 英 [ɪnˈɡriːdiənt] 美 [ɪnˈɡriːdiənt] *n.* 因素；(混合物的) 组成部分
calycosin glucoside　花萼苷
formononetin 英 [ˈfəmɒnəʊnɪtɪn] 美 [ˈfəmɒnoʊnɪtɪn] *n.* 芒柄花黄素

Useful Expressions

1. inconvenient 英 [ˌɪnkənˈviːniənt] 美 [ˌɪnkənˈvinjənt] *adj.* 不便的；不方便的
Can you come at 10:30? I know it's inconvenient for you, but I must see you.
你能 10:30 来吗？我知道你不方便，但是我必须见你。
2. in infancy　婴儿；婴儿期；在婴儿期
Out of five children of two marriages, two died in infancy.

在两次婚姻中生下的五个孩子,有两个在襁褓中就夭折了。

3. adjust 英 [əˈdʒʌst] 美 [əˈdʒʌst] vt. (改变……以) 适应;调整;校正

We have been preparing our fighters to adjust themselves to civil society.

我们一直在培训我们的战士,以使他们适应普通的社会生活。

4. stimulate 英 [ˈstɪmjuleɪt] 美 [ˈstɪmjəˌlet] vt. 刺激;激励;鼓舞;使兴奋

America's priority is rightly to stimulate its economy.

美国的首要任务自然是刺激经济。

Questions

1. What are the difficulties regarding collection of wild midcinal plants?
2. Why should we pay more attention to fertilization in herbal cultivation?
3. Can you tell us how to reduce the damage by pests?
4. What should be taken into account when selecting seeds?
5. What is the difference between herbal cultivation and natural fostering regarding economic input?

5.3 Evaluation of Genetic Diversity of Medicinal Plants

Passage 1

Genetic Diversity in Populations of the Endangered Medicinal Plant Tetrastigma Hemsleyanum Revealed by ISSR and SRAP Markers: Implications for Conservation (Extract)

Materials and Methods

1) Germplasm and DNA Extraction

Because of the narrow distribution, the 15 wild and 12 cultivated accessions were collected from south China, including the only 4 remaining provinces in which T. hemsleyanum is distributed. To obtain more data and increase detecting signal intensity, leaf materials from five plants were collected and mixed in equal proportion as a single individual for DNA extraction, and the above process was repeated three times. Total genomic DNA was extracted following the CTAB method described by Wang et al. DNA quality and quantity were confirmed in 0.8% agarose gels and spectrophotometry. Samples were diluted to 50 ng/uL for PCR amplification.

2) ISSR-PCR

100 anchored microsatellite primers designed by University of British Columbia, Canada, were screened and 11 of them were selected for further study based on their production of reproducible, clear and polymorphic bands. ISSR amplification reactions

were carried out in 25 uL volume containing 2 mM $MgCl_2$, 0.2 mM of each dNTP, 0.3 uM primer, 1.0 U Taq DNA polymerse (Sangon Biotech, China), and 50 ng template DNA. Amplification was performed in an Eppendorf Master Cycler Gradient PCR (Eppendorf, Germany) as follows: initial 5 minutes at 95℃, 35 cycles of 45 seconds at 95℃, 60 seconds at 50℃, 90 seconds at 72℃, and a final 10 minutes extension at 72℃ PCR products were separated on 2% agarose gel and stained by EB and photographed by a Bio-Rad Gel Doc XR + imaging system (Bio-Rad Co., Ltd., USA).

3) SRAP-PCR

8 random SRAP primer combinations were designed and synthesized by Sangon Biotech (Shanghai) Co., Ltd. SRAP amplifications were performed in 25 uL reaction volumes containing 2.5 mM $MgCl_2$, 0.2 mM of each dNTP, 0.3 uM for forward and reverse primer, 1.0 U TaqDNA polymerase (Sangon Biotech, China), and 50 ng genomic DNA. Amplification conditions were as follows, 5 minutes at 95 ℃, 5 cycles of 40 seconds at 94 ℃, 1 minute at 35 ℃, and 1 minute at 72 ℃, and 35 cycles of 40 seconds at 94 ℃, 1 minute at 50 ℃, and 1 minute at 72 ℃, and a final extension of 10 minutes at 72 ℃. PCR products were also analyzed like ISSR analysis.

Data Scoring and Statistical Analysis

The ISSR and SRAP bands were scored as present (1) or absent (0) to form a binary matrix. Dice co-efficient was used to compute a distance matrix between accessions using NTSYS PC2.10 software. Dendrograms were constructed based on Jaccard's similarity coefficients using the UPGMA (unweighted pair-group method with arithmetic means). The Mantel test (MXCOMP in NTSYS, 3,000 permutation) was used to estimate the level of correlation between SRAP and ISSR similarity matrices data. A principal coordinate analysis to construct a three-dimensional array of eigenvectors was performed using the DCENTER module of the NTSYS program. Estimations of within-and between-population variation were obtained with POPGENE software version 1.31 we calculated percentage polymorphic loci (PPB), Nei's gene diversity (H), and Shannon's Information Index (I) of all wild and cultivated accessions. For estimation of inter-population variation, total genetic diversity (Ht), mean intra-population genetic diversity (Hs), the coefficient of genetic differentiation (Gst), and the gene flow (Nm) of the 15 wild accessions were calculated. The correlation between Jaccard's similarity coefficient value and geographical distance between accessions was evaluated using Mantel test (MXCOMP in NTSYS, 3,000 permutation).

Results

1) ISSR Polymorphism

A total of 112 scored bands were amplified using the 11 primers screened across 27 accessions. The number of amplified bands per primer ranged from 17 (U855) to 6 (U822) with an average of about ten bands. Of the 112 bands, 96 (85.71%) were polymorphic, ranging from 13 (U846) to 5 (U827), with an average of 8.73 polymorphic bands per primer. But the polymorphic bands percentage of every geographic population ranged from 11.61% for Guangxi Zhuang Autonomous Region to 42.86% for Zhejiang Province. The genetic diversity of every population differed greatly. Among them, Guangxi Zhuang Autonomout Region and two cultivated populations had quite low value for Shannon's Information Index (I) and Nei's gene diversity (H), while Zhejiang's wild population had the highest Shannon's Information Index (I) for 0.2403 and hn wild population had the highest Nei's gene diversity (H) for 0.1637. Total genetic diversity (Ht), mean within-population genetic diversity (Hs), and the coefficient of Gst of the 4 wild populations yielded 0.2867, 0.1182, and 0.5878, respectively. The results exhibited that 58.78% Gst was among populations and 42.22% was within population. Gene flow among was quite low, with only 0.3507.

2) SRAP Polymorphism

A total of 66 scored bands were amplified using the 8 pairs of primer screened across 27 accessions. The number of bands amplified by each primer combination ranged from 12 (Me5 and Em1) to 6 (Me10 and Em3) with an average of 8.25 bands. Of the 66 bands, 51 (77.27%) were polymorphic, ranging from 10 (Me5 and Em1) to 3 (Me7 and Em3), with an average of 6.37 polymorphic bands per primer. But the polymorphic bands percentage of every geographic population ranged from 6.06% for Tianlin, Guangxi to 30.30% for Hunan Province. The genetic diversity of every population differed greatly. Among them, two cultivated populations had the lowest value for Shannon's Information Index (I) and Nei's gene diversity (H), while Hunan's wild population had the highest Shannon's Information Index (I) for 0.1402 and Jiangxi's wild population had the highest Nei's gene diversity (H) for 0.0909. Total genetic diversity (Ht), mean within-population genetic diversity (Hs), and the coefficient of Gst of the 4 wild populations yielded 0.2407, 0.0731, and 0.6965, respectively. The results exhibited that 69.65% Gst was among populations and 31.35% was within population. Gene flow among populations was only 0.2179.

In general, higher genetic diversity were observed in inter-population than in intra-population, higher genetic diversity were also observed in wild populations than in cultivated populations.

3) Cluster Analysis

The GS (genetic similarities) of wild populations based on the ISSR markers varied from 0.5179 (Zhejiang2 and Guangxi1) to 0.8750 (Zhejiang1 and Zhejiang3) with an average value of 0.610, while the GS of cultivated populations varied from 0.848 (Zhejiang3 and Lishui, Zhejiang6) to 0.991 (Tianlin, Guangxi1 and Tianlin, Guangxi3) with an average value of 0.931. The dendrogram used UPGMA analysis based on Jaccard's GS of the ISSR markers of 15 wild accessions. The 15 wild accessions were clustered into

4 groups at the 0.75 similarity coefficient level. The accessions of all the wild populations from Zhejiang were highly distinct for ISSR polymorphism and form a separate cluster, Cluster II comprised 2 accessions of all the wild populations from Guangxi origin, while cluster III comprised 3 accessions of Jiangxi origin. Cluster IV comprised 5 accessions of Hunan origin. The results of the principal components analysis (PCA) were comparable to the cluster analysis. The first three components explain 61.9% of the total variation.

The GS of wild populations based on the SRAP markers varied from 0.5606 (Zhejiang2 and Guangxi1) to 0.9394 (Zhejiang1 and Zhejiang4) with an average value of 0.731, while the GS of cultivated populations varied from 0.924 (Lishui, Zhejiang2 and Lishui, Zhejiang5) to 1 (Tianlin, Guangxi1 and Tianlin, Guangxi4, Lishui, Zhejiang1 and Lishui, Zhejiang3) with an average value of 0.963. The dendrogram used UPGMA analysis based on Jaccard's GS of the SRAP markers of 15 wild accessions. The 15 wild accessions were also clustered into 4 groups at the 0.73 similarity coefficient level, which was essentially similar to the ISSR markers except the clustering result of some sub-clusters. Groupings identified by UPGMA analysis were confirmed by PCA data which also revealed that Zhejiang populations were genetically very distinct from others' populations. The first three most informative PCA components accounted for 69.6% of the variation observed. The similarity matrices from SRAP and ISSR markers were compared for their correspondence using Mantel's test. The correlation between the coefficient matrices for the SRAP and ISSR data was high ($r=0.869$, $P<0.01$) indicating a great correspondence of polymorphisms brought out by the two marker systems.

In general, the regional character of clustering results was very outstanding, the accessions of the same region tended to be clustered together. Higher level of genetic similarity was observed in intra-population than in inter-population, The average genetic similarity among cultivated populations were far above that among wild populations, which is in accordance with the previous conclusion of genetic diversity.

https://link.springer.com/article/10.1007%2Fs10722-014-0210-6

Vocabulary

genomic 英 [dʒiːˈnəʊmɪk] 美 [dʒiːˈnoʊmɪk] *adj.* 染色体组的

binary matrix   二元矩阵

co-efficient   协同效率

compute 英 [kəmˈpjuːt] 美 [kəmˈpjut] *v.* 计算；估算；用计算机计算

eigenvectors 英 [ˈaɪɡənvektə(r)] 美 [ˈaɪdʒənˌvektə] *n.* 特征向量

polymorphic 英 [ˌpɒlɪˈmɔːfɪk] 美 [ˌpɒlɪˈmɔːfɪk] *adj.* 多形的；多态的

geographic population   地理种群

cluster analysis   聚类分析

a great correspondence of polymorphisms   多态性的极大对应

Useful Expressions

1. photograph 英 [ˈfəʊtəɡrɑːf] 美 [ˈfoʊtəɡræf] *vt.* 为……拍照；拍照；摄影
They were photographed playing with their children.
他们跟孩子一起嬉戏的情景被拍成了照片。
2. random 英 [ˈrændəm] 美 [ˈrændəm] *adj.* 随机的；任意的；胡乱的
The survey used a random sample of two thousand people across England and Wales.
该调查在英格兰和威尔士随机抽样了2000人。
3. cluster 英 [ˈklʌstə(r)] 美 [ˈklʌstɚ] *vt.* 使密集；使聚集
The doctors clustered anxiously around his bed.
医生焦急地围在他的床边。
4. distinct from 不同于……
Engineering and technology are disciplines distinct from one another and from science.
工程学和工艺学互不相同，也有别于自然学科。

Questions

1. How was total genomic DNA of the leaves extracted?
2. What conclusion is reached?
3. What is purpose of SRAP polymorphism?

Passage 2

Genetic Diversity Study of Some Medicinal Plant Accessions Belong to Apiaceae Family Based on Seed Storage Proteins Patterns

Introduction

Apiaceae (Umblliferae) is one of the well-known families among flowering plants because of its worth properties. The plants of Apiaceae are very important in medical, pharmaceutical and chemical industry. They usually are grown in moderate climates and adapted to low amounts of water. It has been showed that there is essence in almost all parts of these plants. Cumin (Cuminum cyminum L.), Fennel (Foeniculum vulgare L.) and Longleaf (Falcaria vulgaris Bernh) are three of the most important medicinal plants belonging to Apiaceae family.

Cumin as a valuable medicinal plant has been used with the people of India and Egypt since ancient times. Fennel is a perennial plant growing wild in many different parts of the world and is so valuable due to its high content of trans-anethole, limonene, estrogele, fenchone, V-terpinene and α-pinene. Falcaria vulgaris (named Paghazeh in Kurdish) grows near farmlands and is very popular as a vegetable in the west of Iran. In the folk medicines, this herb is used for healing of skin wounds and stomach disorders.

Nowadays, wild medicinal plants are really endangered because of inordinate harvest by native people. Therefore, study and evaluation of genetic relationship of these plants is a crucial issue. One of the methods to evaluate genetic diversity of plant and classification of them is the electrophoresis of seed storage proteins. Seed storage proteins are highly independent of environmental fluctuations. Sodium dodecyl sulphate polyacrylamide gel electrophoresis (SDS-PAGE) technique is commonly used for separation of seed storage proteins. The analysis of SDS-PAGE is one of the practical methods to study evolutionary relation of plants. This type of technique has been used to analyze genetic diversity among different species of plants.

This study was conducted in order to evaluate seed protein variability in different Iranian Cumin, Fennel and Longleaf accessions and grouping them based on these proteins as a biochemical marker.

Materials and Methods

This study was performed in Proteomics Laboratory of Medical Biology Research Center, Kermanshah University of Medical Sciences, Kermanshah, Iran. For this research, 29 Cumin, 3 Fennel and 3 Longleaf accessions were collected from different parts of Iran. Then the seeds separately were powdered. The SDS-PAGE and protein assay were made according to Laemmli and Bradford, respectively. The protein extraction carried out with extraction buffer (Tris-HCl, pH 8.5; NP-40, 2%; PMSF, 1 mM as a serine protease inhibitor and EDTA, 1 mM), other steps were done according to Xi et al. SDS-PAGE method in resolving gel with 12.5% acrylamide and stacking gel with 5% acrylamide was applied to assess extracted proteins in these accessions. After electrophoresis, the gel was stained with the Coomassie Brilliant Blue R-250, and destained with methanol and acetic acid. Identifying each protein bond carried out according to standard protein marker contain; ovotransferrin (78 kDa), bovine serum albumin (66 kDa), ovalbumin (45 kDa), actinidin (29 kDa), β-lactoglobulin (18 kDa) and lysozyme (14 kDa).

Following SDS-PAGE bands scoring, the NTSYS-pc software version 2.02 was used for cluster analysis via UPGMA method, principle coordinate analysis (PCoA) and calculating the cophenetic correlation coefficient between the similarity matrix and the cophenetic matrix derived from the dendrogram.

Results and Discussion

The SDS-PAGE of seed storage proteins is a method to investigate protein pattern and classify plant varieties because these proteins are highly preserved. The electrophoresis gels showed 38 bands that indicated low polymorphism among the accessions especially when we considered them separately based on type of plant. For example, polymorphism percentage among Cumin accessions was 0.41 and there was no polymorphism among three Fennel accessions. In similar research and in accordance with this study, Shoorideh showed a low percentage of polymorphism among 24 Cumin accessions (58%).

According to the Jaccard similarity matrix for binary data, the amount of similarity varied from 0.20 to 1.00 (there was no any difference between some accessions). Overall, there was more similarity between Fennel and Cumin ecotypes and it may be because of their seed similarity in size and shape.

Cluster analysis based on binary data, according to UPGMA method classified accession into 3 groups in 0.63 distance unit. The biggest group was the first group which was occupied with all Cumin accessions. The second group consisted of three Fennel ecotypes while last cluster had all three Longleaf accessions in itself. The cophenetic correlation values (0.98) were too high that offered a very good fit of the clustering to the data matrix. In order to establish the relationship among samples and comparison to cluster analysis, principle coordinate analysis was done. Distribution pattern of accessions in this aspect was very mainly similar to the result extracted from cluster analysis.

Conclusion

Results showed that differences in the protein band pattern among the selected plants (Cumin, Fennel and Longleaf) were highly significant but low polymorphism when they considered individually. In addition, this research has shown noticeable similarity and differences between three plants under study based on seed storage proteins pattern. The similarity between Cumin and Fennel accessions was more than between Cumin and Longleaf ones. As a recommend, for demonstration of more differences between accessions based on seed storage proteins pattern, it is better that this research would be done via 2-dimensional electrophoresis in the future.

https://link.springer.com/article/10.1007%2Fs11033-012-1914-3

Vocabulary

Apiaceae   伞形科
worth property   价值财产
cumin 英 ['kʌmɪn] 美 ['kʌmɪn, 'kumɪn, 'kju-] n. 枯茗；蒔蘿；孜然芹；孜然芹果
fennel 英 ['fenl] 美 ['fɛnəl] n. 小茴香；[植]茴香
fluctuation 英 [ˌflʌktʃʊ'eɪʃn] 美 [ˌflʌktʃʊ'eʃən] n. 波动；涨落；[生]彷徨变异
extraction buffer   提取缓冲液；提取液
cophenetic correlation coefficient   共形相关系数
cophenetic matrix   共形矩阵
polymorphism percentage   多态性百分率
binary data   二进制数据；二元数据
cophenetic correlation value   共形相关值

Useful Expressions

1. inordinate 英 [ɪnˈɔːdɪnət] 美 [ɪnˈɔːrdɪnət] *adj.* 无节制的；过度的；紊乱的
They spend an inordinate amount of time talking.
他们花在说上的时间太多了。
2. manifest 英 [ˈmænɪfest] 美 [ˈmænəˌfɛst] *vt.* 显示；表明；证明；使显现
There may be unrecognised cases of manifest injustice of which we are unaware.
也许还有一些我们不知道的明显不公平的现象被忽视了。

Questions

1. Who has used Cumin as a medicinal plant since ancient times?
2. What makes wild medicinal plants endangered?
3. What do the results indicate?

Chapter Six
Conservation and Sustainable Utilization of Medicinal Plant Resources

6.1 Conservation Status of Medicinal Plant Resources

Passage 1

Utilization and Conservation of Medicinal Plants in China (Extract)

Present Status of Utilization of Medicinal Plants in China

The official *Pharmacopoeia of the People's Republic of China* lists 709 different drugs. Among these, only a little more than 40 items are animal and mineral products. Others are all derived from plant materials. Plant materials account for more than 80% of the drugs sold on the market. Approximately 1,000 species of plants are now commonly used in Chinese medicine and about half of these are considered as the main medicinal plants which are in particularly common use. Materials from these 1,000 species are harvested and marketed extensively. Medicinal plants are distributed over a wide geographic area ranging from 50°N (Heilongjiang Province) to 10°N (Hainan Province) and from 80°E to 130°E.

The government has introduced a series of laws, regulations and rules to provide guidelines for and to control the collection, cultivation, production, certification, registration and marketing of medicinal plants in order to promote the development of Chinese medicines in the most appropriate manner and to protect public health and safety.

If we consider the most common ailments affecting wide sections of the population, viz., cancer, cardiovascular diseases, hepatitis, influenza and various kinds of fevers, then the most popularly used medicinal plants are either those in the group used for increasing the person's immunity and resistance to diseases, e.g. Panax notoginseng, Panax quinquefolium, Cordyceps sinensis, Eucommia ulmoides, Gastrodia elata, Ganoderma lucidum, etc., or are among other medicinal plants used for specific conditions, e.g. Dioscorea spp., Salvia miltiorrhiza, Astragalus membranaceus var. mongholicus, Glycyrrhiza uralensis and Bupleurum chinense.

It should be remembered, however, that prescriptions in the Chinese system of medicine nearly always consist of materials from several different plants.

Chapter Six   Conservation and Sustainable Utilization of Medicinal Plant Resources

Trends in the Utilization of Medicinal Plants

Whereas both traditional Chinese medicine and western medicine form part of today's health care system in China, the former is more popular among the country's population at large. The state offers assistance and support to the cultivation of medicinal plants and, at the same time, promotes the development of an integrated health care delivery system which combines traditional Chinese medicine with modern western medicine. Most large hospitals offer treatment based on both systems and it is very common for patients to receive both Chinese medicines and western medicines for their illness. Sometimes they take both kinds of medicines at the same time and sometimes alternatively.

People suffering from ailments for which there is no satisfactory treatment in western medicine, no matter whether they live in rural areas or in cities, always look for alternatives based on traditional Chinese medicine. Many studies comparing the relative efficacies of Chinese and western medicines for a range of conditions, both chronic and acute, have established the superiority of the former. Examples of chronic conditions include chronic anaemia, lupus erythematosus, stomach cancer and ABO haemolytic disease of new-born babies. Examples of acute conditions include mortality from haemorrhagic fever caused by wild mouse, myocardial infarction, pancreatitis, and cholangitis.

Discovery of the anti-malarial drug Qinghaosu from Artemisia annua and the medicine for hepatitis derived from Swertia milensis have further proved the future potential of Chinese medicinal plants. Plants used in traditional Chinese medicine which have shown promise as potential sources of anticancer and anti-HIV compounds include Glycyrrhiza uralensis, Ligusticum chuanxiong, Astragalus membranaceus var. mongholicus, Schisandra chinensis, Atractylodes lancea, Acanthopanax senticosus, Panax ginseng, Lithospermum erythrorrhizon, Senecio scandens and Coptis chinensis.

During the period from 1979 to 1990, forty-two new Chinese medicinal preparations appeared on the market. Among them eleven are for cardiac diseases, five for cancer and six for ailments of the digestive system. The green movement and the current trend of "coming back to the nature" have led to a renewed interest in traditional medicinal plants. A new aspect, namely the use of fresh medicinal plant materials, is worth mentioning. Recent studies carried out at the Institute of Botany, Jiangsu Province and Chinese Academy of Sciences on the efficacy of fresh herbs in treatment for the skin diseases, herpes, caused by viruses. A Chinese medicine named "Eliherpes" has proved particularly effective in the treatment of herpes zoster, herpes simplex and sexual herpes. In a clinical trial involving more than 200 patients, it was found to be effective in 94.36% of the cases. The medicine is inexpensive, safe to use and easy to administer. It provides quick pain relief and rapid healing without sequelae. This new approach using fresh herbs has been developed and applied for treating patients in some hospitals in Nanjing City already.

Overexploitation of Natural Resources and the Necessity for Conservation and Cultivation of Medicinal Plants

Until now, the major part of plant materials used in Chinese medicines have originated from wild sources. Among the 1,000 commonly used medicinal plants, 80% in terms of number species and 60% in terms of total quantity have come from wild sources. Overexploitation is a problem common to all medicinal plants. For most species, the natural reserves are exhausted within 10 to 20 years of collection. For some species the supplies from the wild last only for three or four years, after which no more commercial production is possible. It should be noted that the loss of genetic variation within a given species is usually much more serious and occurs much earlier than the total extinction of the species itself. For example, the content of diosgenin in Dioscorea zingiberensis collected during expeditions in the 1950s averaged only 7% whereas the maximum recorded was about 17%. In materials collected during the 1980s the content was reduced to such an extent that even 4% was considered as high. When the medicine using Swertia milensis was developed for the treatment of hepatitis, a shortage in the supply of plant materials appeared within three to four years.

In the past, the distribution of Glycyrrhiza uralensis was mainly concentrated in Inner Mongolia Autonomous Region. In the 1950s G. uralensis grew over an area of 1.2 million ha. By 1981 this was reduced to only 330 thousand ha. The center of western growing area had therefore moved to the Xinjiang Uygur Autonomous Region. The situation is not any better in the eastern growing area either. In the 1960s the plain of Nenjiang river was famous for its G. uralensis population. Twenty years later there were no bushes with any commercial value left in this area. Preliminary statistics indicate that approximately 6,000 tons of G. uralensis are exported annually and the domestic consumption amounts to several thousand tons.

A similar situation of shortage has also been observed for the following species: Cistanche deserticola, Cordyceps sinensis, Asarum heterotropoides var. mandshuricum, Phellodendron chinense, Eucommia ulmoides, Magnolia officinalis, Gastrodia elata, Ephedra sinica, Acanthopanax senticosus, Bupleurum chinense, Paris polyphylla, Atractylodes lancea and Notopterygium incisum.

The depletion of resources has accelerated with increasing demand for Chinese medicinal preparations both at home and abroad. Since 1979 the demand has increased at a rate of 9% per year. In 1987 the total production had reached 650 thousand tons. There are altogether more than 600 industrial plants producing over 4,000 composite drugs in 40 different formulations. In 1990, the total quantity was 20 thousand tons and the total value RMB 5.5 billion. By 1992, this had increased to RMB 11 billion. Chinese medicinal preparations are exported to more than 100 countries and the total value of exports is in the region of 100—200 million US dollars.

Chapter Six Conservation and Sustainable Utilization of Medicinal Plant Resources

The only solution to this rapid exhaustion of resources is cultivation. Since the 1980s there has been a rapid increase in the area cultivated with medicinal plants. In 1984 the total land area devoted to medicinal plant cultivation was 380 thousand ha which was about 161% of that in 1981.

At present there are more than 250 species of medicinal plants being commercially cultivated. Among them about 60 species have performed particularly well under cultivation. The main crops cultivated are as follows: Gastrodia elata, Panax ginseng, Panax notoginseng, Cordyceps sinensis, Coptis chinensis, Eucommia ulmoides, Glycyrrhiza uralensis, Fritillaria cirrhosa, Fritillaria thunbergii, Astragalus membranaceus var. mongholicus, Asarum heterotropoides var. mandshuricum, Schisandra chinensis, Platycodon grandiflorum, Pinellia ternata, Polygonum multiflorum, Macrocarpium officinale, Dendrobium nobile, Gynostemma pentaphyllum, Saposhnikovia divaricata, Gentiana manshurica, Salvia miltiorrhiza, Anemarrhena asphodeloides, Bupleurum chinense, Corydalis turtschaninovii f. yanhusuo, Crocus sativus, Belamcanda chinensis and Cimicifuga foetida.

Concurrently with these developments in the cultivation of Chinese medicinal plants, many exotic medicinal plants have been introduced and brought under cultivation. About 30 introduced species have successfully been grown during the past 30 years. The main among these are as follows: Eugenia aromatica, Scaphium lychnophorum, Amomum compactum, Strychnos wallichiana, Styrax tonkinensis, Silybum marianum, Panax quinquefolium, Atropa belladonna and Digitalis lanata.

Active research is being carried out on various aspects of cultivation and processing of a wide range of medicinal plants. Examples include rapid propagation of Siraitia grosvenorii and Aloe vera var. chinensis; the use of fermentation technology for the processing of Cordyceps sinensis, Ganoderma lucidum and Armillaria mellea; tissue culture for the production of active principles of Panax ginseng, Panax notoginseng, Ligusticum chuanxiong, Glycyrrhiza uralensis, Panax quinquefolium, Lithospermum erythrorrhizon, Hyoscyamus niger, Corydalis turtschaninovii f. yanhusuo; and imporoving cultural practices for higher yields and better quality of Eucommia ulmoides, Coptis chinensis, Magnolia officinalis, Lonicera japonica and Tripterygium wilfordii.

In addition to the species under cultivation, approximately 2,000 species of exotic medicinal plants are currently maintained in Chinese botanical gardens.

http://www.fao.org/docrep/w7261e/W7261e13.htm

Vocabulary

guideline 英 ['gaɪdˌlaɪn] 美 ['gaɪdˌlaɪn] *n.* 指导方针；指导原则
certification 英 [ˌsɜːtɪfɪ'keɪʃn] 美 [ˌsɜːrtɪfɪ'keɪʃn] *n.* 证明；鉴定；证书
viz. 英 [vɪz] 美 [vɪz] *adv.* <拉>即，就是；亦即

cardiovascular 英 [ˌkɑːdiəʊˈvæskjələ(r)] 美 [ˌkɑːrdioʊˈvæskjələ(r)] *adj.* 心血管的
hepatitis 英 [ˌhepəˈtaɪtɪs] 美 [ˌhɛpəˈtaɪtɪs] *n.* 肝炎
integrated health care delivery system   社区整合性健康照护体系
anaemia 英 [əˈniːmɪə] 美 [əˈnimɪə] *n.* [医]贫血;贫血症
lupus erythematosus   红斑狼疮
haemorrhagic 英 [ˌheməˈrædʒɪk] 美 [ˌheməˈrædʒɪk] *adj.* 出血的
myocardial infarction   心肌梗死
pancreatitis 英 [ˌpænkrɪəˈtaɪtɪs] 美 [ˌpænkrɪrˈtaɪtɪs] *n.* 胰腺炎
cholangitis 英 [kəʊlænˈdʒaɪtɪs] 美 [koʊlænˈdʒaɪtɪs] *n.* 胆管炎
herpes 英 [ˈhɜːpiːz] 美 [ˈhɜːrpiːz] *n.* [医]疱疹
botanical garden   植物园

Useful Expressions

1. acute 英 [əˈkjuːt] 美 [əˈkjut] *adj.* 尖的;锐的;敏感的;剧烈的;[医]急性的
The war has aggravated an acute economic crisis.
战争加剧了原本已很严重的经济危机。
2. accelerate 英 [əkˈseləreɪt] 美 [ækˈsɛləˌret] *vi.* 加快;加速
Growth will accelerate to 2.9 per cent next year.
明年的增长会加快到2.9%。

Questions

1. How many species of main medicinal plants are listed the official *Pharmacopoeia of the People's Republic of China*?
2. What actions does the government take for the development of Chinese medicines?
3. What plant has the anti-malarial drug Qinghaosu?

Passage 2

Role of Biotechnology for Protection of Endangered Medicinal Plants (Extract)

Essence of *in vitro* Culture

Plant cell methods and techniques were initially used in fundamental scientific investigations at the beginning of their development in the early sixties of the last century. Plant biotechnology is based on the totypotence of the plant cell. This process of *de novo* reconstruction of an organism from a cell in differentiated stage is highly linked to the process of dedifferentiation when the cell is returning to its early embryogenic/meristematic stage. In this stage cells undergo division and may form nondifferentiated callus tissue or may redifferentiate to form new tissue, organs and an entire organism. Morphogenesis *in vitro* is realized via two major pathways: (i) Organogenesis when a group of cells is involved for *de novo* formation of organs and (ii) somatic embryogenesis when the new organism is initiated from a single cell.

Micropropagation

Micropropagation is a vegetative propagation of the plants *in vitro* conditions (in glass vessels under controlled conditions) leading to development of numerous plants from the excised tissue and reproducing the genetic potential of the initial donor plant.

Usually tissues containing meristematic cells are used for induction of axilary or adventitious shoots but induction of somatic embryos can be achieved from differentiated cells as well.

Micropropagation is used routinely for many species to obtain a large number of plants with high quality. It is widely applied to agricultural plants, vegetable and ornamental species, and in some less extent to plantation crops. One of the substantial advantages of micropropagation over traditional clonal propagation is the potential of combining rapid large-scale propagation of new genotypes, the use of small amounts of original germplasm (particularly at the early breeding and/or transformation stage, when only a few plants are available), and the generation of pathogen-free propagules. Compared to the other spheres of *in vitro* technologies clonal propagation has proved the greatest economical and market importance in industry including pharmaceutical industry which needs for raw material from the medicinal plants is increasing constantly. It offers faster and alternative way for production of raw material and from another side overcoming the problems arising from the limited natural resources.

At present, there is a long list of research groups worldwide investigating hundreds of medicinal species. Various success procedures and recipes for many of these species have been developed. However, there is not a universal protocol applicable to each species, ecotype, and explant tissue. From another side all these continuous tedious studies on the standardization of explant sources, media composition and physical state, environmental conditions and acclimatization of *in vitro* plants have accumulated information, continuously enriched, which is a good basis for elaboration of successful protocols for more species. Wider practical application of micropropagation depends on reduction of costs so that it can become compatative with seed production or traditional vegetative propagation methods (e. g. cuttings, tubers and bulbs, grafting).

Metabolic Engineering and Biotransformation

The plant cell culture systems have potential for commercial exploitation of secondary metabolites. Similar to the fermentation industry using microorganisms and their enzymes to obtain a desired product plant cells are able to biotransform a suitable substrate compound to the desired product. The latter can be obtained as well by addition of a precursor (a particular compound) into the culture medium of plant cells. In the process of biotransformation, the physicochemical and biological properties of some natural products can be modified. Thus, biotransformation and its ability to release products into the cells or out of them provide an alternative method of supplying valuable natural products that occur in nature at low levels. Generally, the plant products of commercial interest are secondary metabolites, which in turn belong to three main categories: essential oils, glycosides and alkaloids. Plant cell cultures as biotransformation systems have been highlighted for production of pharmaceuticals but other uses have also been suggested as new route for synthesis, for products from plants difficult to grow, or in short supply, as a source of novel chemicals. It is expected that the use, production of market price and structure would bring some of the other compounds to a commercial scale more rapidly and *in vitro* culture products may see further commercialization. The application of molecular biology techniques to produce transgenic cells and to effect the expression and regulation of biosynthetic pathways is also a significant step towards making *in vitro* cultures more generally applicable to the commercial production of secondary metabolites. However, because of the complex and incompletely understood nature of plant cells growing in *in vitro* cultures, case-by-case studies have been used to explain the problems occurring in the production of secondary metabolites from cultured plant cells.

Genetic manipulations (direct and indirect genetic transformation) are other different approaches to increase the content biological active substances in plants. Genetic engineering covers a complex of methods and techniques applied to the genome in order to modify it to obtain cells and organisms with improved qualities or possessing desired traits. These might refer to better yield or resistance, as well as, to higher metabolite production or synthesis of valuable biologically active substances. Gene transfer may be direct when isolated desired DNA fragments are inserted into the cell most often by electrical field or adhesion. This method is less used in medicinal plants. Indirect genetic transformation of plants uses DNA vectors naturally presenting in plant pathogens to transfer the isolated genes of interest and to trigger special metabolic pathways. Agrobacterium rhizogenes induces formation of "hairs" at the roots of dicotyledonous plants. Genetically modified "hairy" roots produce new substances, which very often are in low content. Hairy roots are characterized with genetic stability and are potential highly productive source for valuable secondary metabolites necessary for the pharmaceutical industry. Manipulations and optimization of the productivity of the transformed hairy roots are usually the same as for the other systems for *in vitro* cultivation. They also depend on the species, the ecotype, the explant, the nutrient media, cultivation conditions, etc.

All these applications of the principles of plant cell division and regeneration to practical plant propagation and further manipulations could be possible if there are reliable *in vitro* cultures and their efficiency depends on many various factors.

Factors Influencing Cell Growth *in vitro*

The ability of the plant cell to realize its totypotence is influenced in greatest extend by the genotype, mother/donor plant, explant, and growth regulators what was confirmed by the tedious empirical work of *in vitro* investigations. Here, some of the specific and most important requirements will be mentioned in order to understand the efforts and originality of some ideas when establishing *in vitro* cultures of medicinal plants.

Genotypes: Morphogenetic potential of excised tissue subjected to cultivation *in vitro* is in strong dependence of the genotype. Genetically plants demonstrate different organogenic abilities, which were observed for all plants groups including medicinal plants. Some of the species (like tobacco and carrot) are easy to initiate in *in vitro* cultures while others are more difficult-reculcitrant (cereals, grain legumes, bulbous plants). Many of the wild species like most of the medicinal plants and especially those producing phenols are more difficult or extremely difficult to handle.

Donor plant: The donor plant should be healthy, in the first stage of its intensive growth, not in dormancy. Rhyzomes and bulbs usually need pretreatment with low or high temperatures for different periods of time.

Explant: The explant type might determine the organogenesis potential and the genetical stability of the clonal material. Physiological age of the explant is also crucial. Immature organs and differentiated cells excised from stem tips, axilary buds, embryos and other meristematic tissues are the most appropriate. However, despite the development of cell and molecular biology the limits still exist in receiving easy information about the genetic, epigenetic and physiological status of the explant. Empirical approach is the most common to specify the chemical and physical stimuli triggering cell totypotence.

Nutrient media: Although more than 50 different media formulations have been used for the *in vitro* culture of tissues of various plant species the formulation described by Murashige and Skoog is the most commonly used, often with relatively minor changes. Other famous media are those of Gamborg, Huang and Murashige Nischt, etc. The nutrient medium usually consists of all the essential macro- and micro-salts, vitamins, plant growth regulators, a carbohydrate, and some other organic substances if necessary.

Plant growth regulators: Plant growth regulators, including the phytochormones, are essential for cell dedifferentiation, division and redifferention leading to callus tissue and organ formation. The auxins and cytokinins are the most important for *in vitro* development and morphogenesis. However, the most appropriate plant regulators and their concentrations in the nutrient media depend on the genotype, explants type and the donor plant physiological status. Hence, numerous combinations could be designed and the optimal ones are validated empirically. All that creates the difficulties of the experimental work, which is dedicated to find the balance between the factors determining reliable *in vitro* development.

Cytokinins: Different groups of cytokinins might be used but the most efficient ones for induction of organogenesis and a large number of buds are the natural cytokinins (zeatin and kinetin) or the synthetic ones-6-benzylaminopurine (benzyl adenine (BA, BAP), 6-γ(-dimethylallyl-amino)-purine (2iP) and thidiazuron (TDZ).

Auxins: The auxins also are obtained from natural plant materials like indolyl-3-acetic acid (IAA), indole 3-butyric acid (IBA), α-naphthyl acetic acid (NAA) or are chemically produced like 2,4-dichlorophenoxyacetic acid (2,4-D), 2,4,5-trichlorophenoxyacetic acid (2,4,5-T), picloram, etc. The auxins have a wide spectrum of effects on different processes of plant development and morphogenesis. Depending on their chemical structure and concentration, they induce or inhibit cell division, stimulate callus or root formation.

Gibberellins: The group of gibberellins includes more than 80 compounds, which stimulate cell division and elongation. The most commonly used one is gibberellic acid (GA3).

Vitamins and supplements: Growth regulatory functions are attributed to some of the vitamins B group—thiamine (B1), niacin (vit B3, nicotinic acid, vitamin PP), piridoxin (vit B6), which in fact are the most popular for in vitro recipes. Supplements like yeast extract, coconut milk, maize extract and some other might effect tissue growth and bud development.

The best morphogenesis could be achieved when the optimal balance between the effect of genotype, explant and growth regulators is identified.

https://www.intechopen.com/books/environmental-biotechnology-new-approaches-and-prospective-applications/role-of-biotechnology-for-protection-of-endangered-medicinal-plants

Vocabulary

essence 英 ['esns] 美 ['ɛsəns] *n.* 香精；本质；实质；精华；精髓
in vitro culture   体外培养
de novo   重新；更始
embryogenic 英 ['embrɪəʊ'dʒenɪk] 美 ['embrɪoʊ'dʒenɪk] *adj.* 胚胎发生（形成）的
meristematic 英 [ˌmerɪstə'mætɪk] 美 [ˌmerɪstə'mætɪk] *adj.* 分生组织的
callus tissue   愈伤组织
morphogenesis in vitro   体外形态发生
via 英 ['vaɪə] 美 ['vaɪə, 'viə] *prep.* 经过；通过；凭借；取道
micro propagation   微繁殖
vegetative propagation   营养（体）繁殖；无性繁殖
axillary 英 ['æksɪləri] 美 ['æksɪləri] *adj.* 腋窝的；叶腋的；腋生的
adventitious 英 [ˌædven'tɪʃəs] 美 [ˌædven'tɪʃəs, -vən-] *adj.* 外来的；[生]偶生的；后天的

recipe 英 ['resəpi] 美 ['resəpi] *n.* 食谱；处方；秘诀
protocol 英 ['prəʊtəkɒl] 美 ['proʊtəkɔːl] *n.* 科学实验报告（或计划）
applicable 英 [ə'plɪkəbl] 美 ['æplɪkəbəl, ə'plɪkə-] *adj.* 适当的；可应用的
tedious 英 ['tiːdiəs] 美 ['tidiəs] *adj.* 冗长乏味的；单调沉闷的；令人生厌的
acclimatization 英 [əˌklaɪmətaɪ'zeɪʃn] 美 [əˌklaɪmətɪ'zeɪʃn] *n.* 驯化；环境适应性
elaboration 英 [ɪˌlæbə'reɪʃn] 美 [ɪˌlæbə'reʃən] *n.* 精致；精心制作（或计划）；详尽（阐述）
biotransformation 英 [baɪətrænsfə'meɪʃn] 美 [baɪətrænsfə'meɪʃn] *n.* 生物转化
fermentation 英 [ˌfɜːmen'teɪʃn] 美 [ˌfəːmən'teʃən, -mɛn-] *n.* 发酵；激动；纷扰
microorganism 英 [ˌmaɪkrəʊ'ɔːgənɪzəm] 美 [ˌmaɪkro'ɔrgəˌnɪzəm] *n.* 微生物
enzyme 英 ['enzaɪm] 美 ['ɛnzaɪm] *n.* [生化]酶
precursor 英 [priː'kɜːsə(r)] 美 [priː'kɜːrsə(r)] *n.* 前辈；先锋；预兆；初期形式
glycoside 英 ['glaɪkəˌsaɪd] 美 ['glaɪkəˌsaɪd] *n.* 配糖；配糖类
genome 英 ['dʒiːnəʊm] 美 ['dʒiːnoʊm] *n.* 基因组；染色体组
dicotyledonous 英 ['daɪkɒtɪliːdənəs] 美 ['daɪkɒtɪliːdənəs] *adj.* 双子叶的

Useful Expressions

1. substantial 英 [səb'stænʃl] 美 [səb'stænʃəl] *adj.* 大量的；牢固的；重大的
The party has just lost office and with it a substantial number of seats.
该党刚刚竞选失利，同时还失去了许多席位。

2. substrate 英 ['sʌbstreɪt] 美 ['sʌbˌstret] *n.* 底物；基底；底层；基层
The features and microstructure of the cladding layer and interface layer between cladding layer and copper substrate were analyzed.
人们对熔覆层和熔覆层与基体结合区的形貌及微观组织进行了研究。

3. empirical 英 [ɪm'pɪrɪkl] 美 [ɛm'pɪrɪkəl] *adj.* 凭经验的；以观察或实验为依据的
There is no empirical evidence to support his thesis.
他的论文缺乏实验证据的支持。

Questions

1. What is micropropagation?
2. What is the significance of metabolic engineering and biotransformation?
3. What influences cell growth in vitro?

6.2 Conservation Strategies of Medicinal Plant Resources

Passage 1

Evaluation of Medicinal Plant Resources and Strategies for Conservation (Extract)

Intrdouction

Since the first earth summit in Rio de Janeiro, there has been a sustained global awareness of the importance of the superfluity of biodiversity and natural resources from tropical forests for several purposes. This stems not only from the derivable forest products, but also from the potent ethno-botanical and ethno-medicinal uses of the plants in these forests. The world's tropical rain forests are especially rich in biodiversity but there is rapid depletion of these natural resources in Nigeria and possibly worldwide. These pressures which arise from degradation, unsustainable arable land use, urbanization and industrialization are taking their toll as well. The plant genetic resources of Nigeria, are a veritable source of pharmaceuticals and therapeutics. Traditional medicine practice has existed in Africa and other cultures for centuries since man came into being but until recently, has been neglected or even outlawed in some cases due to undue pressure from practitioners of modern medical practice and the unscientific background of its method of operation. Okujagu opined that this worldwide renewed interest in traditional medicine derived from the realization that: In the absence of widespread use of modern or orthodox medicine in these poor countries, healthcare has virtually been sustained by these cultural alternatives. Ayodele had challenged Nigerian taxonomists and conservation biologists to undertake proper identification and conservation of these highly important genetic resources. According to WHO in 2001, 80% of the world population, relied almost exclusively on traditional medicines using natural substances mostly derived from plants in the treatment of diseases.

Medicinal plants have been defined as any plant with one or more of its organs containing substances that can be used for therapeutic purpose or which can be used as precursors for the synthesis of antimicrobial drugs. The need to study medicinal plants cannot be overemphasized for a number of reasons including, (i) widespread use of plants in folk medicine, (ii) rescuing traditional medicinal plants and knowledge about them from imminent loss as well as (iii) their role in attaining the health for all target. More so, a clear understanding of the biological data of an organism or species provides an insight on the potentials inherent in it to solve problems such as potentials of plant extracts in controlling different plant diseases. Today, many medicinal plants face extinction with about 15,000 medicinal plants under threat and for most of the endangered species no conservation action has been taken.

All over the world studies of the vegetation of different areas are undertaken to document the flora, especially in these days of remarkable genetic losses due to over exploitation of forests and its products. Apart from this, some of these workers have undertaken their studies to ascertain the different medicinal plants in their regions. A lot of work has been undertaken to ascertain and document the medicinal properties of indigenous plants in Nigeria. Among the pioneer workers were Bhat et al. who undertook an ethnobotanical study in Kwara State, Nigeria to document the uses of the different plant resources by the people. From interviewing the elderly and traditional medicine practitioners, they listed the different plants, their botanical names, parts used, method of preparation, ailments used for as well as local names in the three major Nigerian languages. Among the plants they studied were Daniellia oliveri, Jatropha curcas, Azadirachta indica, Blumea gariepina, Calotropis procera, Citrus aurantifolia, Bryophyllum pinnatum, Bridelia ferruginea, Argemone mexicana, Glypheae brevis, Vitellaria paradoxa, Eleusine indica, Ficus thonningii, Cassia occidentalis, Cassia alata, Hyptis suaveolens, Carica papaya among others. Iwu carried out an intensive inventory of medicinal plant resources of West and Central Africa for the United States Army medicinal research and material command. Sofowora in his book *Medicinal Plants and Traditional Medicine in Africa* had listed numerous medicinal plants in Africa, and Nigeria in particular. He enumerated the medicinal plants in common use, some research publications in this area and biologically active ingredients derived from some of these plants.

Nwosu had undertaken an ethnobotanical survey of southern Nigerian pteridophytes. The survey which involved several field trips between January 1996, to May 1999 revealed an interesting diversity and distribution of ferns and fern allied plants. She recorded 36 pteridophyte species belonging to 22 families which were used by the locals for medicines, food, cosmestics, fodder, aphrodisiacs and for manure. Some of the commonest species were Adiantum, Asplenium, Arthromeris, Aleuriopteris, Lycopodium, Lygodium, Dryopteris, Ophioglossum, Pteris and Selaginella species.

Ibe and Nwufo using questionnaires, personal interviews and review of available records documented that in southeastern Nigeria medicinal plants also served as vegetables, fruits, trees and ornamentals. Moreover, out of forty-three plants they reported that about fifteen were undergoing domestication. Their study revealed that much has not been done as regards the domestication of medicinal plants in southeastern Nigeria. Soladoye et al. studied the angiosperm community in the permanent site of Olabisi Onabanjo University, Ago Iwoye, Ogun State, Nigeria with the aim of conserving them for posterity especially during the development of the new university campus. They listed one hundred and thirty-eight species belonging to fifty-five families. There were one hundred and twenty-seven dicotyledonous species while monocots were eleven. Leguminosae appeared the dominant family followed successively by Rubiaceae and Euphorbiaceae. Furthermore, they encountered fifty-four trees, forty-three shrubs, ten climbers, twenty-eight herbs and three grasses/sedges.

Ogbole et al. also worked in five local governments in Ogun State investigating the different plants used in treating inflammatory diseases using semi structured questionnaires administered to traditional medical practitioners, herbalists and herb sellers. Among the species they enumerated were also some of those encountered in the Redemption City survey. Common species to both studies included Alstonia boonei, Vernonia amygdalina, Citrus aurantifolia, Cassia fistula, Alternanthera sessilis, Alchornea cordifolia, Combretum zenkeri, Carica papaya, Xylopia aethiopica, Dioclea reflexa, Citrus limon, Corchorus olitorius, Ananas comosus, Celosia argentia, Aframomum melegueta, Piper guineensis, Cocos nucifera, Vigna unguiculata and Abrus precatorius.

Sofidiya et al. had carried out a survey on the different plants used as anti-inflammatory agents in herb markets in Lagos. They recorded forty-one plants belonging to twenty-three families through direct interview with traditional herb sellers. They outlined the botanical and local names of each, together with the part used, method of preparation and administration. The most commonly used parts of the plants were the leaves.

Ndukwu and Ben-Nwadibia worked on the ethnomedicinal potentials of different plants used as spices and condiments in the Niger Delta. They had listed 24 species cutting across ten plant families among which were Dennettia tripetala, Xylopia aethiopica, Ocimum americanus, O. gratissimum, O. canum, O. basilicum, O. viride, Myristica fragrans, Piper guineensis, P. nigrum, P. umbellatum, Capsicum annuum, C. frutescens, C. minimum, Aframomum melegueta, Zingiber officinale, Murraya koenigii, Allium cepa, A. sativum, Thymus vulgaris and Pergularia daemia.

Again, Okujagu et al. following the establishment of the Nigeria Natural Medicine Development Agency undertook a survey of medicinal plants in the southeastern, northeastern, north central and southwestern parts of Nigeria through questionnaires and interviews of both scientists and traditional medicine practitioners.

Furthermore, Odugbemi undertook a comprehensive study of medicinal plants in Nigeria highlighting their names, botanical families, distribution, their uses, parts used, mode of preparation. He also reviewed works done by different researchers on some of these medicinal plants. He had listed 831 medicinal plant species with illustrations which were very useful in their collection and identification. Most of his data was got through interviews of traditional healers. He challenged scientists to undertake quality researches to validate the claims of the traditional healers on the uses of these plants.

Chapter Six Conservation and Sustainable Utilization of Medicinal Plant Resources

Ubom documented the ethnobotanical and biodiversity conservation of plant resources in the Niger Delta and listed about three hundred and thirty-nine plant species which were used by the Niger Delta dwellers for purposes such as medicinal, food/condiments, fuel, commercial uses (fruits, beverages, timber, spices, thickeners, etc.). The commonest among these were Acanthus montanus, Anthocleita vogelii, A. djalonensis, Antiaris africana, Alstonia boonei, A. congoensis, Elaeis guinensis, Raphia hookeri, Dacryodes edulis, Cocos nucifera, Irvingia gabonensis, Hevea brasiliensis, Pterocarpus santalinoides, Lonchocarpus cayanescens, Milicia excelsa, Daniellia ogea, Newbouldia laevis, Napoleoona vogelii, Mimosa pigra, Nauclea diderrichii, Musanga cecropioides, Paullinia pinnata and a host of others.

From the foregoing, a lot of researches have been going on to ascertain whether the claims of the traditional medicine healers are true and more work still have to be done in order to extract the active ingredients contained in these plants and to promote the need for their conservation through awareness programmes in villages, seminars, workshops, etc. This study would document the plant resources available for research, so that both researchers and students will have a better understanding of plant biodiversity, conservation practices and the different medicinal plants within their reach.

https://scialert.net/fulltext/?doi=jbs.2012.34.42

Vocabulary

superfluity 英 [ˌsjuːpəˈfluːəti] 美 [ˌsjupəˈfluɪti] *n.* 过剩；过多；多余物
ethno-botanical   民族植物学
ethno-medicinal   民族药
veritable 英 [ˈverɪtəbl] 美 [ˈvɛrɪtəbəl] *adj.* 名副其实的；真正的
therapeutics 英 [ˌθerəˈpjuːtɪks] 美 [ˌθɛrəˈpjutɪks] *n.* 治疗学；疗法
taxonomist 英 [tækˈsɒnəmɪst] 美 [tækˈsɑːnəmɪst] *n.* 分类学者
synthesis of antimicrobial drugs   抗菌药物的合成
enumerate 英 [ɪˈnjuːməreɪt] 美 [ɪˈnuːməreɪt] *vt.* 列举；枚举；数
domestication 英 [dəˌmestɪˈkeɪʃn] 美 [dəˌmɛstəˈkeʃən] *n.* 驯养；驯化
posterity 英 [pɒˈsterəti] 美 [pɑːˈsterəti] *n.* 子孙；后裔；后代；儿孙
climber 英 [ˈklaɪmə(r)] 美 [ˈklaɪmɚ] *n.* 登山者；[植]攀缘植物；[动]攀禽类
sedge 英 [sedʒ] 美 [sɛdʒ] *n.* [植]莎草
forego 英 [fɔːˈɡəʊ] 美 [fɔrˈgo, for-] *vt.* 放弃；摒弃；摒绝
active ingredient   活性组分；有效成分

Useful Expressions

1. outlaw 英 [ˈaʊtlɔː] 美 [ˈaʊtˌlɔ] *vt.* 宣布……为不合法；将……放逐；剥夺……的法律保护

The regional governor has been given powers to outlaw strikes and expel suspected troublemakers.

地方长官被赋予禁止罢工、驱逐嫌疑闹事者的权力。

2. define 英 [dɪˈfaɪn] 美 [dɪˈfaɪn] vt. 规定；使明确；精确地解释；画出……的线条

We were unable to define what exactly was wrong with him.

我们说不清楚他到底哪里不对劲。

3. imminent 英 [ˈɪmɪnənt] 美 [ˈɪmɪnənt] adj. 迫在眉睫的；迫切的；危急的；逼近的

There appeared no imminent danger.

眼前似乎没有危险。

4. ascertain 英 [ˌæsəˈteɪn] 美 [ˌæsərˈteɪn] vt. 弄清；确定；查明

Through doing this, the teacher will be able to ascertain the extent to which the child understands what he is reading.

这样一来，老师就能确定孩子们对他所读的内容理解多少了。

Questions

1. What is medicinal plants defined?
2. What did Sofidiya and his colleagues find from their survey on the different plants?
3. Why do researchers undertake so many studies on medicinal plants?

Passage 2

Medicinal Plant Conservation

Medicinal plant conservation strategies need to be understood and planned for based on an understanding of indigenous knowledge and practices. Many drugs contain herbal ingredients, and it has been said that 70%—80% of the world's population rely on some form of non-conventional medicine and around 25%—40% of all prescription drugs contain active ingredients derived from plants in the United States alone. Many countries rely on these medicinal plants for the health and well being of its population, but the market demand has led to an increased pressure on the natural resources that lend to the production of some of these plants. The most serious proximate threats when extracting medicinal plants generally are habitat loss, habitat degradation, and over harvesting. Developing markets for natural products, particularly those that are harvested from the wild, can trigger a demand that cannot be met by available or legal supplies and demands a conservation initiative so the local populations are not exploited, causing more damage to their resources. Many times populations are taken advantage of for their resources and knowledge, which can often be used for financial gain. Conservation of medicinal plants in its biocultural perspective not only implies conservation of biodiversity, but also places an

Chapter Six Conservation and Sustainable Utilization of Medicinal Plant Resources

equal emphasis on conservation of cultural diversity.

Around the Globe

Asia represents one of the most important centers of knowledge with regard to the use of plant species for treatment of various diseases. Kunwar states, it has been estimated that the Himalayan region harbors over 10,000 species of medicinal and aromatic plants, supporting the livelihoods of about 600 million people living in the area. In Nepal they use a traditional healing system that is called Ayurveda, which is influenced by Buddhism and Hinduism's central ideas of balance in life. High-altitude medicinal plants provide quality products, and this is the reason why they are often the first choice of local users as immediate therapy and by pharmaceutical companies as precious ingredients. When it comes to profits made in the communities up to 50% of the Nepal's rural household's income, they're derived from commercial collection of medicinal and aromatic plants.

KwaZulu-Natal, South Africa community of Mnoqobokazi has high unemployment rates in the area and reliance on subsistence agriculture and wild produce. Socioeconomic factors such as low education levels and lack of access to western health care have been cited as important reasons for reliance on indigenous medicine in South Africa. Both villagers and healers in the area would cultivate one or more species because they could only be found far away, or were frequently used, or had to be fresh when used, or they were planted as protection against witchcraft. Also there were ten people from Mnqobokazi, mainly women, who harvested plants on a commercial basis to conserve. An interview with a conservation officer at the Wetland Park claimed that harvesting of medicinal plants was not a problem in this part as it was further north. In other parts of South Africa the most frequently used medicinal plants are slow-growing forest trees, in which the bark and underground parts are mainly the parts utilized. Because there is a high demand for such resources, the trees are becoming endangered and a lot of the collection is unrestricted. Regulations are now being placed on some of the resources that originally had been exploited and many schools and research facilities are working together to come up with new ways to foster their beloved trees and still manage to get what is needed from the trees as well by proposing the idea of substituting the bark or underground parts with leaves of the same plant.

Samoa has had a great influence on western medicine when it comes to finding a cure for HIV/AIDS. New research has shown that the isolation of prostratin, found in the bark of the Samoan mamala tree, from Homalanthus nutans has led to the extreme potency against HIV-1. Both the National Cancer Institute and Brigham Young University have guaranteed to return to the Samoan people a significant portion of any royalties. Paul Alan Cox, an American ethnobotanist, raised money based on awareness of environmental degradation due to logging, in order to protect the 30,000-acre (120 km^2) lowland forest of Falealupo village on the island of Savaii. The Swedish Society for the Conservation of

Nature established three new indigenously controlled preserves. Controlled preserves cause controversy because in traditional Polynesian societies, land, including the natural plant and animal populations, which occupied it, were viewed as sacred and an ancestral inheritance. Western approaches to conservation on indigenous land and within an indigenous community must collaborate and understand indigenous knowledge systems in order to conserve cultural identity. Paul Cox stated that the loss of these indigenous knowledge systems may yet prove to be one of the greatest tragedies of our age. The US National Park Service officials, the American Samoan Government, and the traditional chiefs and orators of the villages of Tafua had agreed to lease their lands for 50 years to the U.S National Park Service in order to protect American Samoa's rain forests. The Tafua Rainforest Preserves received funding from the Swedish International Development Authority, which was used to secure water supply, improve roads, and offer assistance in the development of village-based environmental tourism in Tafua. Cox explains, "All parties to these agreements agree that any development of tourism must be village initiatives, rather than foreign initiatives, and must be carefully planned and controlled so that the Samoan culture in these areas is not jeopardized."

http://en.academic.ru/dic.nsf/enwiki/11720121/

Vocabulary

non-conventional 英 [ˌnɒnkənv'enʃənl] 美 [ˌnɒnkənv'enʃənl] *adj.* 非传统的
prescription drug　须医师处方才可买的药品
habitat degradation　生境退化
conservation initiative　保护倡议
ayurveda 英 [eɪɜː'viːdə] 美 [eɪɜː'viːdə] *n.* 印度草医学
witchcraft 英 ['wɪtʃkrɑːft] 美 ['wɪtʃkræft] *n.* 巫术；魔法
isolation of prostratin　前列腺素的分离
mamala tree　马马拉树
inheritance 英 [ɪn'herɪtəns] 美 [ɪn'hɛrɪtəns] *n.* 继承；遗传；遗产

Useful Expressions

1. proximate 英 ['prɒksɪmət] 美 ['prɑːksɪmət] *adj.*（时间、顺序等方面）近邻的；最接近的
The proximate causes of this recovery are traditional.
本次复苏的"近因"是传统性的。
2. controversy 英 ['kɒntrəvɜːsi] 美 ['kɑːntrəvɜːrsi] *n.* 论战；公开辩论
The proposed cuts have caused considerable controversy.
削减开支的提议引起了诸多争议。

Questions

1. What does the market demand do to the natural resources of medicinal plants?
2. Where is the center of knowledge with regard to the use of plant species for treatment of various diseases?
3. What is the disadvantage of controlled preserves?

6.3 Main Ways and Methods of Protecting Medicinal Plant Resources

Passage 1

Ecological Protection of Medicinal Woody Plants (Abstract)

Medicinal woody plants, especially medicinal tall trees, play the same important role in forest structure, ecological balance and timber production as other tree species in forest, and due to their additional medicinal values overuse of these trees is more intensive than others. Many medicinal materials are destructively obtained from plants such as roots or bark used as medicinal materials. The contradiction between the utilization and protection of medicinal woody plants becomes more and more incisive. Based on the analysis of the utilized situation and specialty of medicinal woody plants, the trouble between the plants protection and utilization was observed, the method to solve it and the fundamental research work were discussed. The following aspects of researches were suggested to be conducted (i) study on the distribution in organs, seasonal and age variations, and correlation with environmental factors of principal medicinal compositions in mature trees to clear the optimum of harvest and cultivation conditions; (ii) study on the distribution in organs, seasonal and age variations, and correlation with environmental factors of principal medicinal compositions in saplings, especially the time course of the variation in medicinal compositions and biomass to achieve the optimal tree ages for the balance between biomass and production of medicinal products during saplings development; (iii) study on the influence and regulation of environmental factors on medicinal compounds production in woody plants to look for the optimal cultivated conditions for optimizing the accumulation of biomass and medicinal chemicals; (iv) further study on the regulatory mechanism of the induced production of main medicinal compositions by ecological factors at protein (key enzyme) and gene level to accumulate fundamental data for the enhancement of quality and quantity, and approach of new accesses to medicinal products using biological technology (cell culture and gene technology). Aimed at medicinal woody plants in Chinese forest resources, to develop the fundamental researches on resources protection and rational utilization will create many profound scientific significances. Firstly, medicinal woody plants are the important components of Chinese natural forest resources, so the problem for their protection and utilization, especially for that of tall trees, is quite remarkable and special. To reveal the internal contradictory between plant resources protection and its reasonable exploitation and exploit a practicable access to solve it will promote the fulfillment of natural forest protection project in China. Secondly, traditional Chinese medicine is a main part of Chinese excellent ancestral culture, and the traditional utilizing models have been carried on for thousands of years. Accompanying with the development of human society, many unavoidable troubles such as the shortage of natural resources and the pollution of natural environment are more and more severely, which makes the old models of the traditional Chinese medicine become more and more harmful and inaccessible to mankind. New substitutive approach to the utilization of traditional Chinese medicine, especially to that of Chinese medicinal woody plants will be one of the key methods to improve the present situation. Thirdly, traditional Chinese medicine, the cherish treasure of Chinese ancestral culture, needs not only to be preserved but also to be developed. One of the main problems to restrict the extensive spread of the traditional Chinese medicine is its unstable quality, therefore, to reach the stable and good quality is tightly linked to the improvement of traditional Chinese medicine. Hence the environmental regulation to the cultivation of medicinal plants, which can prove and guarantee the stable and good quality, will fit the demand on the production of medicinal plant materials, and correspond to the goal of great efficacy and superior quality during the course of modernization of traditional Chinese medicine.

https://eurekamag.com/research/048/855/048855375.php

Vocabulary

woody 英 ['wʊdi] 美 ['wʊdi] *adj.* 木质的；木本的
optimum 英 ['ɒptɪməm] 美 ['ɑːptɪməm] *n.* 最佳效果；[生] 最适度
sapling 英 ['sæplɪŋ] 美 ['sæplɪŋ] *n.* 幼树；树苗
biomass 英 ['baɪəʊmæs] 美 ['baɪoʊmæs] *n.* （单位面积或体积内）生物的数量
ancestral culture   祖先文化
the cherish treasure   珍宝

Useful Expressions

1. intensive 英 [ɪn'tensɪv] 美 [ɪn'tensɪv] *adj.* 加强的；强烈的；[农] 精耕细作的
Each counsellor undergoes an intensive training programme before beginning work.
每个辅导员在上岗前都要接受密集培训。
2. incisive 英 [ɪn'saɪsɪv] 美 [ɪn'saɪsɪv] *adj.* 尖锐的；深刻的；直接的
She's incredibly incisive, incredibly intelligent.

她机敏过人,才智超群。

3. contradictory 英 [ˌkɒntrəˈdɪktəri] 美 [ˌkɑːntrəˈdɪktəri] adj. 矛盾的;反驳的;抗辩的
Customs officials have made a series of contradictory statements about the equipment.
海关官员们对这种设备做出了一系列互相矛盾的陈述。

4. accompany with 伴随着;兼带着
The lightnig is accompanied with thunder.
电闪雷鸣。

Questions

1. Why are medicinal woody plant more overused?

2. What makes the old models of the traditional Chinese medicine become more and more harmful?

3. What can we do to prove and guarantee the stable and high quality of Chinese medicine?

Passage 2

Protecting Traditional Medicinal Knowledge in Zimbabwe (Extract)

The African Potato (Hypoxis Hemeracallidae) was hot in Zimbabwe in 2000. For years it had appeared in some of the street markets in the capital of Harare alongside other muti, like ginger, as a remedy for stomach aches. Who first began marketing it for its "cure-all" effects cannot be known at this late date because all of the local muti merchants started selling it when they saw its increased popularity with Zimbabweans.

At one point, however, being accountable to no one other than impersonal Zimbabwean consumers, the muti merchants began to display another tuber they also called the African Potato. That potato did not possess any of the active healing ingredients of the original. Accusations of fraud and concerns for public safety followed, forcing the president of the Zimbabwe National Association of Traditional Healers Gordon Chavanduka to appear on national television to distinguish between the bogus potato and the curative African Potato. Although Chavanduka did not reveal the full secret of what the African Potato cured, how to prepare it, or in what dosage it was safe to take it, it was evident to most that the potato he identified as the "real" African Potato was powerful. Hence, as a caveat, Chavanduka had to insist that consumers of the African Potato should always consult a traditional healer for their health needs, so they could be assured they were given the right plant.

Tradition Facing Outside Pressures

Despite intense interest from pharmaceutical companies, drug legislation agencies, and ethnobotanical sciences in identifying the "right plants" for health and other uses from developing countries, the story of the African Potato reveals that an additional element is significant in preserving biodiversity: identifying the "right individuals" and the "right practices" attached to such plants. In Zimbabwe in particular, individual practitioners provide services based on traditional medicinal knowledge (TMK) of local plants and do not traditionally seek either profit or patent for these medicinal plants' utility. But because of the weight of international and national agendas that favor novelty and commercial application in intellectual property rights (IPR) legislation, these local individuals may soon be divested of the TMK they maintain and make valuable.

The agenda looming most directly over the future of traditional medicinal practitioners in Zimbabwe is seen in Article 27.3(b) on Trade-Related Intellectual Property Rights (TRIPS) of the World Trade Organization (WTO). While IPR has historically applied only to works of art and/or industrial innovations, the TRIPS article extends exclusive ownership rights to include the products and processes derived from biological substances. During the WTO Seattle Meeting in 2000 it was this extension of IPR that became the crux of a split between biodiversity-rich countries wishing to maintain sovereignty over their cultural and biological resources and the industrial countries who were already developing and profiting from those resources. Under Article 27.3 (c), however, developing countries and the African Union (AU), including member-nation Zimbabwe, are allowed to develop their unique and separate (Sui generis) protection frameworks that would mediate their comparative advantage in biological diversity. Yet the capacity of individual practitioners to continue servicing their local communities through their cultural practices may remain unmediated and unprotected even in these planned national Sui generis legislations. This is the case especially because both the AU and Zimbabwe's Sui generis would subsume these individuals into their own state agendas for technological and economic development.

In Zimbabwe this situation will need special attention because two-thirds of the population rely on traditional healers for some portions of their health needs. Indeed, in the rural areas where two-thirds of Zimbabweans live, the healer continues to create and use cultural resources like Bantu languages (Shona, Ndebele, and Tonga) and customs for the local community, which participates, understands, and confirms these resources further. Subsequently, local and agrarian needs and ways of life need to be considered separately from the national agendas to develop the natural substances of TMK for foreign export and/or for metropolitan benefit and development.

TMK Practitioners

A spectrum of individuals use and share TMK in Zimbabwe. However, not all of them need or deserve special protection. This is because not all TMK practitioners follow the traditional customs of accessing, using, sharing, keeping, and valuing TMK as a symbolic resource with regard to the needs of the surrounding community, or totem (mutopa) group.

In fact, the only way to gain intimate or first-hand knowledge of TMK is through an intense mentorship with an elder healer, called the *n'anga*, wherein children, usually kin, are initiated as assistants, called *makumbi*. During their initiation time, it aquires ritual practices related to collecting, preparing, delivering, narrating, and keeping biological substances that have healing power through the healer's oral tradition and their own hands-on imitation. The early induction into these customary practices is significant cognitively, because the Shona people believe younger individuals are closer to the spirit of the ancestors and are therefore more likely to access and receive their knowledge of them in dreams. The practical side of initiation involves introduction to the physical elements of TMK that can be any of the parts of plant roots, leaves, bark, stems, fruits, grasses, aloes, seeds, thorns, climbers, and symbiotic insect-plant growths, as well as any part of an animal and its excretions.

The actual Shona term used to refer to these substances is *mishonga*—which translates to mean both "magic" and "medicine" simultaneously. Further, as several anthropologists have concluded, the truth and effectiveness of these natural substances can be drawn out and confirmed in ritual narratives that artfully dance a fine line between a plausible fiction and an unlikely fact. In these rituals, the healer/practitioner wields both natural substances and symbols that allude to powerful historical, ancestral, spiritual, and physical valences. While the symbols and the actually natural substrates are potentially the heritage of all Zimbabweans, the accumulated intuition (or art) in wielding them in distinctive and effective ways is part of the personal magic that each healer can claim as trademark. Indeed, while the actually natural substances, the material culture, and information about medicinal plants properties may be taken away by outsiders. "The TMK practices cannot be applied for," says Sub-Headman of Manjolo Communal Lands, Aident Majiki.

TMK Practices

People cannot apply to learn TMK as if applying for a class and TMK cannot be taken away; customary practitioners literally embody it in their practices, making it inalienable. The actual keeping of this knowledge by a new practitioner begins typically only after the *n'anga* has died and there is no one else to guide them with the customary practices of TMK anymore. At this point the apprentice undergoes a special process called *kusvikiro*, dreaming while receiving specific instruction and guidance from the ancestor spirits. In many cases, the spirit that guides apprentices in these dreams is the elder healers themselves. The knowledge new healers possess seems to be distinct, latent, and personal—similar to a talent or gift. Indeed, one who has a spirit (shave) on them is viewed with special respect under customary codes. These individuals are also tested by community authorities in their ability to channel the ancestors' reservoir of culture. "There is a part of African culture that is understood but also a part that no one can know unless you are a *mudzimu* (ancestor)," says Clayton Gungawo, a Zimbabwean *mbira* (thumb piano) player. The Shona firmly believe that only those healers who have access to the ancestors also have access to the full repertoire of TMK.

https://www.culturalsurvival.org/publications/cultural-survival-quarterly/protecting-traditional-medicinal-knowledge-zimbabwe

Vocabulary

merchant 英 [ˈmɜːtʃənt] 美 [ˈmɜːrtʃənt] *n.* 商人；店主；批发商；零售商
impersonal 英 [ɪmˈpɜːsənl] 美 [ɪmˈpɜːrsənl] *adj.* 没有人情味的；非个人的
tuber 英 [ˈtjuːbə(r)] 美 [ˈtuːbə(r)] *n.* 结节；(植物的) 块茎
fraud 英 [frɔːd] 美 [frɔd] *n.* 欺诈；骗子；伪劣品；冒牌货
bogus potato 假马铃薯
caveat 英 [ˈkæviæt] 美 [ˈkeviæt] *n.* 警告；附加说明
agenda 英 [əˈdʒendə] 美 [əˈdʒendə] *n.* 议事日程；日常工作事项
crux 英 [krʌks] 美 [krʌks, krʊks] *n.* 症结；关键
split 英 [splɪt] 美 [splɪt] *vt.* 分裂
sovereignty 英 [ˈsɒvrənti] 美 [ˈsɑːvrənti] *n.* 主权国家；国家的主权；独立自主
subsume 英 [səbˈsjuːm] 美 [səbˈsuːm] *vt.* 归入；包括
agrarian 英 [əˈɡreəriən] 美 [əˈɡrerien] *adj.* 土地的；农业的
totem 英 [ˈtəʊtəm] 美 [ˈtoʊtəm] *n.* 图腾；图腾形象；崇拜物
intimate or first-hand knowledge 亲近或第一手的知识
mentorship 英 [ˈmentəʃɪp] 美 [ˈmentəʃɪp] *n.* 制度；师徒制；导师制
kin 英 [kɪn] 美 [kɪn] *n.* 亲戚；家族；门第；亲属关系
initiation time 开始时间
simultaneously 英 [ˌsɪməlˈteɪniəsli] 美 [ˌsaɪməlˈteɪniəsli] *adv.* 同时地
anthropologist 英 [ˌænθrəˈpɒlədʒɪst] 美 [ˌænθrəˈpɑːlədʒɪst] *n.* 人类学家
valence 英 [ˈveɪləns] 美 [ˈveɪləns] *n.* (化合) 价；原子价
wield 英 [wiːld] 美 [wild] *vt.* 行使；使用 (武器、工具等)
apprentice 英 [əˈprentɪs] 美 [əˈprentɪs] *n.* 学徒；徒弟；新手
latent 英 [ˈleɪtnt] 美 [ˈletnt] *adj.* 潜在的；潜伏的；休眠的
reservoir of culture 文化贮存库
repertoire 英 [ˈrepətwɑː(r)] 美 [ˈrepərtwɑː(r)] *n.* 全部节目；全部本领

Useful Expressions

1. accountable 英 [əˈkaʊntəbl] 美 [əˈkaʊntəbəl] *adj.* 负有责任的
Public officials can finally be held accountable for their actions.

终于要对政府官员实行问责了。

2. loom 英 [lu:m] 美 [lum] *vi.* 朦胧出现

Vincent loomed over me, as pale and grey as a tombstone.

文森特赫然出现在我面前,他的面色灰白得像一块墓碑。

3. induction 英 [ɪnˈdʌkʃn] 美 [ɪnˈdʌkʃən] *n.* 诱发;归纳(法)

Every induction is a speculation.

所有归纳推理都是一种猜测。

4. cognitively 英 [ˈkɑːgnətɪvli] 美 [ˈkɑːgnətɪvli] *adv.* 认知地

It is where you cognitively decide that you are going to work towards a goal and achieve that goal without being sidetracked by "instant" distractions.

那是你凭感知决定你将要朝着一个目标努力并达到那个目标,而不会被即刻的分心的事物导入旁轨的状况。

5. have access to 使用;接近;可以利用

They now have access to the mass markets of Japan and the UK.

他们现在进入了日本和英国的大众市场。

Questions

1. Why was the African Potato popular in Zimbabwe in 2000?
2. Does the African Potato have any healing effects claimed by the merchants?
3. What does the story of the African Potato tell us?
4. Why are children initiated as assistants in ritual practices?
5. How are apprentices trained?

6.4　Protection of Rare and Endangered Medicinal Plant Resources

Passage 1

Endangered Species

The environmental movement reached its peak with the enactment of the Endangered Species Act (ESA) of 1973. As public concern over environmental degradation heightened, Congress passed the most sweeping piece of environmental legislation in American history. When President Richard M. Nixon signed the law on 28 December 1973, he enthusiastically proclaimed that nothing was more priceless and more worthy of preservation than the wildlife with which the country had been blessed. Intent on fulfilling Nixon's mandate, the authors of the Endangered Species Act made an unmistakably strong statement on national species protection policy. The ESA provided for the protection of ecosystems, the conservation of endangered and threatened species, and the enforcement of all treaties related to wildlife preservation.

Pre-ESA Protection Efforts

Endangered species existed long before 1973, of course. The protection of individual species was an incremental process. Rooted in the tradition of colonial law, U. S. Supreme Court decisions through the nineteenth century ensured state jurisdictional control over that of landowners. By the 1870s, the federal government made it clear that it had an interest in wildlife issues. The establishment of the U. S. Fish Commission in 1871 and Yellowstone National Park in 1872 increased the role of the federal government substantially. The tension between federal and state authority resulted in the Yellowstone Game Protection Act of 1894, which established Yellowstone as a *de facto* national wildlife refuge in order to protect bison.

In the first half of the twentieth century, the federal government increased its direct, national jurisdiction with such legislation as the Lacey Act (1900), the creation of the first official national wildlife refuge at Pelican Island (1903), the ratification of the Migratory Bird Treaty Act with Canada (1918), and the passage of the Bald Eagle Protection Act (1940). Yet, a comprehensive national policy on species preservation was not enacted until the 1960s. The professionalization of ecology and the dawning of the American environmental movement created the needed atmosphere for reform. Building on the political response to Rachel Carson's *Silent Spring* (1962), the Bureau of Sport Fisheries and Wildlife established the Committee on Rare and Endangered Wildlife Species in 1964. The committee of nine biologists published a prototypical list of wildlife in danger of extinction, entitled the *Redbook*, listing sixty-three endangered species. Congress passed a more comprehensive Endangered Species Preservation Act in 1966, requiring all federal agencies to prohibit the taking of endangered species on national wildlife refuges and authorizing additional refuges for conservation. The follow-up Endangered Species Conservation Act of 1969 extended protection to invertebrates. It also expanded prohibitions on interstate commerce provided by the Lacey Act and called for the development of a list of globally endangered species by the secretary of the Interior. The directive to facilitate an international conservation effort resulted in the Convention on International Trade in Endangered Species of Wild Fauna and Flora in early 1973. This set the stage for the Endangered Species Act later that year.

Passage of ESA and Early Challenges

Despite a surge of environmental regulatory lawmaking in the early 1970s, including the Clean Air Act, Federal Water Pollution Control Act Amendments (Clean Water Act), Federal Environmental Pesticide Control Act, and Coastal Zone Management Act, debate continued regarding.

Federal and state regulatory authority and the types of species warranting protection.

Chapter Six Conservation and Sustainable Utilization of Medicinal Plant Resources

Representative John Dingell, who introduced the bill that became the Endangered Species Act, insisted that all flora and fauna be included. Section 29 (a) of the ESA makes this clear by stating that all "species of fish, wildlife, and plants are of aesthetic, ecological, educational, historical, recreational, and scientific value to the Nation and its people". The issue of regulation resulted in greater compromise. Section 6, which directs the secretary of the Interior to foster cooperative agreements with states while allowing them substantial involvement in species management, also provides funds for state programs. In an effort to address these issues and others, including the geographical extent of prohibitions and the location of governmental responsibility, the House worked on fourteen different versions while the Senate worked on three. The bill was ultimately passed by both houses of Congress almost unanimously, setting a clear mandate (with only twelve dissenting votes in the House and one in the Senate). The subsequent history of ESA was much more highly contested.

One of the first major challenges to the ESA came with the Tennessee Valley Authority (TVA) v. Hill battle over the Tellico Dam. From its inception, the Tellico Dam project of TVA faced major challenges. In the early 1970s, a lawsuit charging the violation of the 1969 National Environmental Policy Act (NEPA) and an inadequate environmental impact statement delayed construction. Resuming construction in 1973, the project halted again in 1977 when a lawsuit charged Tellico with violating the Endangered Species Act. The discovery of a small fish, the snail darter, in the portion of the Little Tennessee River yet to be swallowed up by the dam, created what later became a textbook case in environmental ethics. U. S. Attorney General Griffin Bell, who argued the TVA case himself, compared the three-inch fish to the social and economic welfare of countless people. The Supreme Court's response was unequivocal. With the law upheld, the project stopped in its tracks. When the ESA subsequently came up for reauthorization in 1978, a plan to provide a mechanism for dispute resolution, in cases like Tellico, resulted in the creation of the first major change in ESA. The Endangered Species Committee, dubbed the "God Squad", was given the power to decide when economic and societal interests outweighed the biological consequences. Ironically, after the committee rejected the exemption for Tellico, populations of snail darters were found in neighboring Tennessee creeks. This discovery came after the authorization for Tellico's completion squeaked through in an amendment to the 1979 Energy and Water Development Appropriations Act.

While the "God Squad" had refused the exemption for Tellico, the committee opened the door for mitigation plans by considering "alternative habitats" for endangered species. An exemption granted in 1979 to the Grayrocks Dam and Reservoir in Wyoming, which threatened whooping crane habitat downstream, became the precursor to the Habitat Conservation Plan (HCP). A 1982 amendment to ESA created HCPs as an effort to resolve alleged unequal treatment in federal and private sectors. HCPs allowed for the incidental taking of endangered species by private property owners in exchange for the creation of a plan to offset losses through separate conservation efforts. By 1990, the US Fish and Wildlife Service (FWS) had formally approved seven HCPs, with twenty more under way.

Struggles Between Competing Interests in the 1990s

The final extended reauthorization of ESA in 1988 allotted appropriations for five years. Amendments provided funding for state cooperative programs, encouraged the use of emergency powers to list backlogged species candidates, and strengthened the protection of endangered plants. Since 1993, however, Congress has authorized funds only in one-year increments, while bills to weaken ESA have been regularly introduced. The apparent ambivalence with respect to reauthorization reflected divisions between protagonists and antagonists for a strengthened ESA. Conservation organizations such as the World Wildlife Fund and the Nature Conservancy, along with activist oriented organizations such as the Sierra Club and the National Wildlife Federation, grew in strength and numbers during the 1990s, while demanding an expanded ESA. Meanwhile, private property advocates represented by the loose-knit but widespread "wise use" movement led efforts to stop ESA intrusion into the lives of private landowners. The National Endangered Species Act Reform Coalition was particularly effective at getting legislation introduced to modify ESA.

The widely publicized controversy over the northern spotted owl epitomized the struggle of competing interests. The US Forest Service and the Bureau of Land Management advocated protection of this Pacific Northwest subspecies as early as 1977. Yet, the FWS listed the owl as threatened species thirteen years later, in 1990, after years of recommendations for habitat preservation by scientific and environmental coalitions. The "God Squad" met for the third time in fourteen years, in 1993, to discuss the northern spotted owl. Amidst emotional media coverage of the plight of loggers and their families, thirteen out of forty-four tracts of land were opened up, as environmental regulations like ESA took the blame for contributing to economic hardship. While environmentalists used the spotted owl as a surrogate for old growth forests, the timber industry criticized the use of the owl to protect old growth trees. A resolution ultimately took the intervention of President Bill Clinton. The president organized a "Forest Summit" in 1993 to develop the Pacific Northwest Plan, which included a substantial reduction in timber harvesting, an ecosystem-based management plan for 25 million acres of federal land, and an economic plan for displaced loggers and their families.

The Pacific Northwest Plan signaled a shift in federal endangered species policy. In 1995 the National Research Council report on the ESA argued that an ecosystem-based approach to managing natural resources must maintain biological diversity before individual species are in dire trouble. The Clinton administration's Interagency Ecosystem

Management Task Force echoed this proactive approach in their 1995 report, which called for a collaboratively developed vision of desired future conditions that integrated ecological, economic, and social factors.

The shift toward an ecosystem approach follows historical changes in the primary cause of species endangerment from overharvesting to habitat destruction to ecosystem-wide degradation. The history of ESA demonstrates that competing economic goals, political priorities, and ethical arguments have also made solutions more elusive.

https://www.encyclopedia.com/science-and-technology/biology-and-genetics/environmental-studies/endangered-species

Vocabulary

enactment 英 [ɪ'næktmənt] 美 [ɛn'æktmənt] n. 制定；颁布；法律；法规
mandate 英 ['mændeɪt] 美 ['mæn,det] n. 授权；命令；委任；任期
incremental 英 [ˌɪŋkrə'mentl] 美 [ˌɪŋkrə'mentl] adj. 增加的
jurisdictional 英 [ˌdʒʊərɪs'dɪkʃn] 美 [ˌdʒʊrɪs'dɪkʃn] adj. 管辖权的；司法权的
substantially 英 [səb'stænʃəli] 美 [səb'stænʃəli] adv. 大体上；实质上；充分地
a de facto national wildlife refuge 阿德国家野生动物保护区
bison 英 ['baɪsn] 美 ['baɪsən, -zən] n. 野牛
jurisdiction 英 [ˌdʒʊərɪs'dɪkʃn] 美 [ˌdʒʊrɪs'dɪkʃn] n. 管辖权；管辖范围
invertebrate 英 [ɪn'vɜːtɪbrət] 美 [ɪn'vɜːrtɪbrət] n. 无脊椎动物
aesthetic 英 [iːs'θetɪk] 美 [ɛs'θetɪk] adj. 审美的；美的；美学的
inception 英 [ɪn'sepʃn] 美 [ɪn'sɛpʃən] n. 开始；开端；初期
ethics 英 ['eθɪks] 美 ['eθɪks] n. 伦理学；行为准则
unequivocal 英 [ˌʌnɪ'kwɪvəkl] 美 [ˌʌnɪ'kwɪvəkəl] adj. 明确的；毫不含糊的
exemption 英 [ɪg'zempʃn] 美 [ɪg'zempʃən] n. （义务等的）免除；免（税）
offset 英 ['ɒfset] 美 ['ɔːfset] n. 开端；出发
allot 英 [ə'lɒt] 美 [ə'lɑːt] vt. 拨给；分配；摊派给
backlog 英 ['bæklɒg] 美 ['bæklɔːg] vt. (使)积压；储存
increment 英 ['ɪŋkrəmənt] 美 ['ɪnkrəmənt, 'ɪŋ-] n. 增量；增长；增额
ambivalence 英 [æm'bɪvələns] 美 [æm'bɪvələns] n. 矛盾心理；摇摆；犹豫
protagonist 英 [prə'tægənɪst] 美 [pro'tægənɪst] n. 主角；主人公；主要参与者
antagonist 英 [æn'tægənɪst] 美 [æn'tægənɪst] n. 敌手；对抗药
coalition 英 [ˌkəʊə'lɪʃn] 美 [ˌkoʊə'lɪʃn] n. 联合；同盟；结合体
plight 英 [plaɪt] 美 [plaɪt] n. 境况；困境；誓约
logger 英 ['lɒgə(r)] 美 ['lɔːgə(r)] n. 记录器；樵夫；伐木工
tract 英 [trækt] 美 [trækt] n. 大片土地
ethical arguments 道德的争论

Useful Expressions

1. enact 英 [ɪˈnækt] 美 [ɛnˈækt] vt. 制定法律；规定；颁布；颁布
The authorities have failed so far to enact a law allowing unrestricted emigration.
到目前为止，当局未能通过允许自由移民的法律。

2. surge 英 [sɜːdʒ] 美 [sɜːrdʒ] v. 汹涌；起大浪；蜂拥而来
Specialists see various reasons for the recent surge in inflation.
专家们认为造成目前通货膨胀加剧的原因有多种。

3. amendment 英 [əˈmendmənt] 美 [əˈmɛndmənt] n. 修正案；修改；修订
Parliament gained certain rights of amendment.
议会得到了一定的修正权利。

4. allege 英 [əˈledʒ] 美 [əˈlɛdʒ] vt. 断言；宣称；辩解
She alleged that there was rampant drug use among the male members of the group.
她声称该团体中有大量的男性成员吸毒。

5. epitomize 英 [ɪˈpɪtəmaɪz] 美 [ɪˈpɪtəˌmaɪz] vt. 概括；成为……的缩影
Lyonnais cooking is epitomized by the so-called "bouchons".
被称作"bouchons"的里昂小餐馆是里昂饮食的缩影。

Questions

1. What is most precious in the view of President Richard M. Nixon?
2. What did Representative John Dingell do to protect the endangered species?
3. What is the "God Squad"?

Passage 2

Strategies to Protect Rare Australian Plants

Some of Australia's plant species are naturally rare, while others have become rare due to road building, land clearing for homes, businesses and other human influence. The number of rare and threatened Australian plants has increased in the past 100 years and to include approximately 23 percent of its native plants. The government and conservation organizations are working together to limit the number of species that might become extinct.

Assessment

In Queensland, the Department of Environment and Resource Management has studied plants in the wild and determined that 23 species are extinct, 151 are endangered, 274 are vulnerable and 688 are rare. It concluded that all studied species might become extinct in less than 50 years if proactive measures are not undertaken to help these plants come back from the brink of extinction. Other Australian states, such as New South Wales, have also been active in assessing the rare plants within the borders. More than 600 plant species have been assessed as "under threat of extinction" in this Australian state. One of the methods being used to ensure that the rare plants of New South Wales will not be lost is a seed bank, in association with England's Kew Gardens.

Research and Education

The Australian Centre for Plant Biodiversity Research is at the forefront of research efforts in its region. Research that scientists conduct is vital to understanding ways to conserve species of rare and endangered plants in Australia and around the world. Comparisons of genetic variation between rare species and similar, more common species, are some of the methods scientists use to help them understand the plants and develop guidelines for managing and conserving the rare species. Governmental agencies such as the Queensland Department of Environment and Resource Management are actively informing the public about Australia's rare plants and educational efforts are in place to advise people to purchase native plants from nurseries rather than collecting them from the wild.

Conservation and Cultivation

Plant nurseries and botanical gardens have become instrumental in conserving Australia's rare plants because they propagate and cultivate these species in order to conserve the plants found in the wild. In Queensland, the Parks and Wildlife Service has a nursery where workers cultivate more than 50 rare species. Some of these plants have been reintroduced into the wild in an effort to rehabilitate native plant populations.

Habitat Protection

National parks are one of the strategies Australia has implemented to protect its rare plants. Other protected areas that exist will also contribute greatly to saving these special plants by creating areas where no collection is allowed and no development or land clearing will take place. The Society for Growing Australian Plants cultivates many native plants and botanic gardens throughout Australia is also instrumental in growing these plants in protected areas.

Legislation

Many plants are collected for eventual sale in the plant trade, both in Australia and other countries. Many species, such as tree ferns and orchids, have been seriously depleted because of over collection. Common plants such as Australia's grasstrees and staghorn ferns are impacted and might join the list of already rare species in the future. The Australian government has instituted a licensing system for those who collect plants from the wild. It restricts the collection of desirable native plants and those that are classified as rare or endangered. The collector of such plants is required to tag them to identify the exact species and where it originated. Legislation continues, in an attempt to increase the controls on collecting and selling Australia's rare plants.

https://www.gardenguides.com/122836-strategies-protect-rare-australian-plants.html

Vocabulary

assessment 英 [ə'sesmənt] 美 [ə'sɛsmənt] *n.* 评估；评价
proactive 英 [ˌprəʊ'æktɪv] 美 [ˌproʊ'æktɪv] *adj.* 积极主动的
forefront 英 ['fɔːfrʌnt] 美 ['fɔːrfrʌnt] *n.* 前列；最前部；活动中心
legislation 英 [ˌledʒɪs'leɪʃn] 美 [ˌlɛdʒɪ'sleʃən] *n.* 立法；法律；法规
fern 英 [fɜːn] 美 [fɜːrn] *n.* [植]羊齿植物；蕨类植物
grasstree 英 ['ɡrɑːs.tri] 美 ['ɡræs.tri] *n.* 草树
staghorn 英 ['stæɡhɔːn] 美 ['stæɡhɔːn] *n.* 雄鹿角；石松；鹿角大珊瑚
restrict 英 [rɪ'strɪkt] 美 [rɪ'strɪkt] *vt.* 限制；限定；约束；束缚
tag 英 [tæɡ] 美 [tæɡ] *n.* 标签；附属物

Useful Expressions

1. brink 英 [brɪŋk] 美 [brɪŋk] *n.* 边缘；初始状态
Their economy is teetering on the brink of collapse.
他们的经济徘徊在崩溃的边缘。
2. conserve 英 [kən'sɜːv] 美 [kən'sɜːrv] *vt.* 保护；保存；[化,物] 使守恒
The Republic's factories have closed for the weekend to conserve energy.
为了节约能源，该共和国的工厂周末不开工。
3. deplete 英 [dɪ'pliːt] 美 [dɪ'plit] *vt.* 耗尽；用尽；使枯竭
They fired in long bursts, which depleted their ammunition.
他们长时间开火，耗尽了弹药。

Questions

1. What causes the rarity of Australia's plant species?
2. How many species are endangered in Queensland?
3. What is the function of national parks in Australia?

6.5 International Conventions, Policies and Regulations Concerning the Protection of Biological Resources

Passage 1

Convention on Biological Diversity (Exctract)

The Convention on Biological Diversity (CBD), known informally as the Biodiversity

Convention, is a multilateral treaty. The Convention has three main goals including the conservation of biological diversity (or biodiversity); the sustainable use of its components; and the fair and equitable sharing of benefits arising from genetic resources.

In other words, its objective is to develop national strategies for the conservation and sustainable use of biological diversity. It is often seen as the key document regarding sustainable development. The Convention was opened for signature at the Earth Summit in Rio de Janeiro on 5 June, 1992 and entered into force on 29 December, 1993. At the 2010 10th Conference of Parties (COP) to the Convention on Biological Diversity in October in Nagoya, Japan, the Nagoya Protocol was adopted.

Origin and Scope

The notion of an international convention on biological diversity was conceived at a United Nations Environment Programme (UNEP) Ad Hoc Working Group of Experts on Biological Diversity in November 1988. The subsequent year, the Ad Hoc Working Group of Technical and Legal Experts was established for the drafting of a legal text which addressed the conservation and sustainable use of biological diversity, as well as the sharing of benefits arising from their utilization with sovereign states and local communities. In 1991, an intergovernmental negotiating committee was established, tasked with finalizing the convention's text.

A conference for the Adoption of the Agreed Text of the Convention on Biological Diversity was held in Nairobi, Kenya, in 1992, and its conclusions were distilled in the Nairobi Final Act. The Convention's text was opened for signature on June 5, 1992 at the United Nations Conference on Environment and Development. By its closing date, 4 June, 1993, the convention had received 168 signatures. It entered into force on 29 December, 1993.

The convention recognizes for the first time in international law that the conservation of biological conservation of biodiversity is "a common concern of humankind" and is an integral part of the development process. The agreement covers all ecosystems, species, and genetic resources. It links traditional conservation efforts to the economic goal of using biological resources sustainably. It sets principles for the fair and equitable sharing of the benefits arising from the use of genetic resources, notably those destined for commercial use. It also covers the rapidly expanding field of biotechnology through its Cartagena Protocol on Biosafety, addressing technology development and transfer, benefit-sharing and biosafety issues. Importantly, the Convention is legally binding; countries that join it ("Parties") are obliged to implement its provisions.

The convention reminds decision-makers that natural resources are not infinite and sets out a philosophy of sustainable use. While past conservation efforts were aimed at protecting particular species and habitats, the convention recognizes that ecosystems, species and genes must be used for the benefit of humans. However, this should be done in a way and at a rate that does not lead to the long-term decline of biological diversity.

The convention also offers decision-makers guidance based on the precautionary principle which demands that where there is a threat of significant reduction or loss of biological diversity, lack of full scientific certainty should not be used as a reason for postponing measures to avoid or minimize such a threat. The convention acknowledges that substantial investments are required to conserve biological diversity. It argues, however, that conservation will bring us significant environmental, economic and social benefits in return.

The Convention on Biological Diversity of 2010 banned some forms of geoengineering.

Issues

Some of the many issues dealt with under the convention include the following.

Measures, the incentives for the conservation and sustainable use of biological diversity.

Regulated access to genetic resources and traditional knowledge, including Prior Informed Consent of the party providing resources.

Sharing, in a fair and equitable way, the results of research and development and the benefits arising from the commercial and other utilization of genetic resources with the contracting party providing such resources (governments and/or local communities that provided the traditional knowledge or biodiversity resources utilized).

The Biosafety Protocol makes clear that products from new technologies must be based on the precautionary principle and allow developing nations to balance public health against economic benefits. For example, it will let countries ban imports of a genetically modified organism if they feel there is not enough scientific evidence to prove that the product is safe and require exporters to label shipments containing genetically modified commodities such as corn or cotton.

The required number of 50 instruments of ratification/accession/approval/acceptance by countries was reached in May 2003. In accordance with the provisions of its Article 37, the Protocol entered into force on 11 September, 2003.

Global Strategy for Plant Conservation

In April 2002, the parties of the UN CBD adopted the recommendations of the Gran Canaria Declaration Calling for a Global Plant Conservation Strategy, and adopted a 16-point plan aiming to slow the rate of plant extinctions around the world by 2010.

Parties

As of 2016, the convention has 196 parties, including 195 states and the European Union. All UN member states—with the exception of the United States—had ratified the treaty. Non-UN member states that had ratified were the Cook Islands, Niue, and the State of Palestine. The Holy See and the states with limited recognition were non-parties. The US signed but not ratified the treaty, and didn't announce plans to ratify it.

Chapter Six Conservation and Sustainable Utilization of Medicinal Plant Resources

International Bodies Established

Conference of the parties: The convention's governing body is the Conference of the Parties (COP), consisting of all governments (and regional economic integration organizations) that have ratified the treaty. This ultimate authority reviews progress under the convention, identifies new priorities, and sets work plans for members. The COP can also make amendments to the convention, create expert advisory bodies, review progress reports by member nations, and collaborate with other international organizations and agreements.

The Conference of the Parties uses expertise and support from several other bodies that are established by the convention. In addition to committees or mechanisms established on an ad hoc basis, two main organs are as follows.

Secretariat: The CBD Secretariat, based in Montreal, operates under the United Nations Environment Programme. Its main functions are to organize meetings, draft documents, assist member governments in the implementation of the programme of work, coordinate with other international organizations, and collect and disseminate information.

Subsidiary Body on Scientific, Technical and Technological Advice (SBSTTA): The SBSTTA is a committee composed of experts from member governments competent in relevant fields. It plays a key role in making recommendations to the COP on scientific and technical issues. The 13th Meeting of the Subsidiary Body on Scientific, Technical and Technological Advice (SBSTTA-13) held from 18 to 22 February 2008 in the Food and Agriculture Organization at Rome, Italy. SBSTTA-13 delegates met in the Committee of the Whole in the morning to finalize and adopt recommendations on the in-depth reviews of the work programmes on agricultural and forest biodiversity and SBSTTA's modus operandi for the consideration of new and emerging issues. The closing plenary convened in the afternoon to adopt recommendations on inland waters biodiversity, marine biodiversity, invasive alien species and biodiversity and climate change. The current chairperson of the SBSTTA is Dr. Senka Barudanovic.

https://en.wikipedia.org/wiki/Convention_on_Biological_Diversity

Vocabulary

convention 英 [kən'venʃn] 美 [kən'vɛnʃən] n. 会议；国际公约
intergovernmental 英 [ˌɪntəgʌvən'mentl] 美 [ˌɪntərˌgʌvərn'mentl] adj. 政府间的
distill 英 [dɪs'tɪl] 美 [dɪ'stɪl] v. 蒸馏；提取；滴下
biosafety 英 [biːəʊ'seɪftɪ] 美 [biːoʊ'seɪftɪ] n. 生物研究安全性
Cartagena Protocol on Biosafety 《卡塔赫纳生物安全议定书》
modify 英 ['mɒdɪfaɪ] 美 ['mɑːdɪfaɪ] vi. 修改；被修饰
ratification 英 [ˌrætɪfɪ'keɪʃn] 美 [ˌrætɪfɪ'keɪʃn] n. 正式批准；认可；承认
ratify 英 ['rætɪfaɪ] 美 ['rætəˌfaɪ] v. 批准；认可

treaty 英['tri:ti] 美['triti] n. 条约；协议；协商；谈判
ad hoc 特设的；特别的；临时的
disseminate 英[dɪ'semɪneɪt] 美[dɪ'sɛmə,net] vt. 散布；传播
subsidiary 英[səb'sɪdiəri] 美[səb'sɪdieri] adj. 附带的；补足的
modus operandi 做法；惯技
plenary 英['pli:nəri] 美['plinəri, 'plɛnə-] adj. 全体出席的；无限的

Useful Expressions

1. plenary 英['pli:nəri] 美['plinəri, 'plɛnə-] adj. 无限的；完全的；绝对的
The programme was approved at a plenary session of the Central Committee last week.
这个方案在上周的中央委员会全体会议上获得了通过。

2. equitable 英['ekwɪtəbl] 美['ɛkwɪtəbəl] adj. 合理；公平的；平衡法的
He has urged them to come to an equitable compromise that gives Hughes his proper due.
他已经催促他们达成合理的妥协，给予休斯他所应得的报酬。

3. destine 英['destɪn] 美['dɛstɪn] vt. 注定；预定；命定；指定
See come to destiny to destine me to have no perfect world.
看来命运注定了我没有完美的世界。

4. collaborate 英[kə'læbəreɪt] 美[kə'læbə,ret] vi. 合作；协作
He collaborated with his son Michael on the English translation of a text on food production.
他和儿子迈克尔合作，把一篇介绍食品生产的文章译成了英文。

5. convene 英[kən'vi:n] 美[kən'vin] vt. 召集；聚集；传唤
Last August he convened a meeting of his closest advisers at Camp David.
去年8月，他召集自己最亲近的顾问在戴维营开会。

Questions

1. What are the three main objectives of the Convention?
2. How many signatures had the Convention obtained by its closing date?
3. What does the convention cover?

Passage 2

Protecting Biological Resources——Implementation of the Nagoya Protocol in Europe

Introduction

There is no doubt that the world's diverse biological resources are vital for social and economic development. Accordingly, the cosmetics and pharmaceutical markets are no strangers to nature-based research and introducing active ingredients in their products that were originally extracted from some rare plants, fungus or other microorganism, situated in a remote part of the world.

With the growing recognition of the value of such assets, the growing commitment to sustainable development and the need to safeguard the interests of indigenous and local communities providing these assets or contributing their traditional knowledge associated with these assets, attempts have been made over time to give the providers (often located in developing countries) the authority to control access to natural resources. In return, the providers receive a share of the benefits and rewards reaped by the organisations (often based in developed countries) that are developing and commercialising these unique biological assets. Of note is the International Convention on Biological Diversity (the Convention) which was adopted in 1992 to provide a legal framework for access to genetic resources (from such assets) and for sharing any benefits (monetary or otherwise) arising from the utilisation of these genetic resources with the states providing these resources. The follow-on Nagoya Protocol (the Protocol) was adopted on 29 October, 2010 by the signatories to the Convention in an attempt to clarify the Convention so that the signatories could implement the Convention into their national law.

The European Union and its member states are parties to the Convention and, therefore, the new EU Regulation No. 511/2014 (the Regulation) brings EU law in line with the framework agreed at Nagoya. This Regulation will apply once the Protocol enters into force for the EU on 12 October, 2014. Given its potential impact on research (academic or otherwise), some breathing space has been provided, such that some of the key provisions of these regulations will not apply until one year after the Regulation comes into force.

Rationale

The objectives of these various legal instruments are (i) to ensure the conservation of biological diversity and the sustainable use of its components—to achieve this, the countries providing genetic resources and any traditional knowledge associated with such genetic resources (also referred to as "associated traditional knowledge") have been given the right to control access to the resources and to the traditional knowledge; and (ii) to ensure that the providers receive fair and equitable compensation or benefit if genetic resources are exploited, particularly on a commercial basis. The Convention refers to this exploitation as "utilisation" of the genetic resources, and the Protocol and now the Regulation defines "utilisation" as "to conduct research and development on the genetic and/or biochemical composition of genetic resources".

Scope

The Regulation applies to "genetic resources" which it defines as "genetic material of actual or potential value". "Genetic material" is "any material of plant, animal, microbial or other origin containing functional units of heredity". "Associated traditional knowledge" is defined as "the traditional knowledge held by an indigenous or local community that is relevant for the utilisation of genetic resources and that is as such described in the mutually agreed terms applying to the utilisation of genetic resources".

Excluded from the scope of the Regulation are human genetic resources and genetic resources obtained from beyond national jurisdictions (for example, from the high seas). The Regulation also makes it clear that, in a manner similar to the Protocol, the Regulation does not apply to genetic resources, where access is governed by other international instruments. For example, the Regulation does not apply or affect the sharing of influenza viruses of human pandemic potential and access to vaccines for such strains of influenza, as such sharing and access are governed by the pandemic influenza preparedness framework (PIP Framework).

The Regulation obliges users (a natural or a legal person) of the genetic resources and/or the associated traditional knowledge ("Users") to exercise due diligence to ascertain that the genetic resources and any associated traditional knowledge are being accessed in accordance with applicable legislation or regulatory requirements, and that benefits are fairly shared upon mutually agreed terms. This due diligence obligation applies to all Users irrespective of size. Notable Users under this Regulation would be the food and feed industry, the pharmaceutical and cosmetics industry, and academic researchers.

In order to comply with this obligation, Users are required to seek and keep (for 20 years) and to require any subsequent Users to seek and keep an internationally recognised certificate of compliance. Where such certificate is not available, Users must seek and keep the following:

(i) the date and place of access to genetic resources or any associated traditional knowledge;

(ii) the description of utilised genetic resources or any associated traditional knowledge;

(iii) the source from which the genetic resources or any associated traditional knowledge were obtained;

(iv) the presence or absence of rights and obligations regarding access and benefiting sharing;

(v) access permits, where applicable; and

(vi) any additional terms, if any that were agreed between Users and the provider of the genetic resources and/or the associated traditional knowledge.

The Regulation recognises that there may be some uncertainty amongst Users as to the exact measures required to comply with this obligation and, therefore, provides for associations of Users to submit their notion of what constitutes best practice for recognition by the Commission. The Regulation encourages Users to be guided by such accepted best practices.

Monitoring Compliance by Users

Users' compliance will, in the first instance, be monitored by competent authorities appointed by member states.

At identified points in the chain of activities which constitute utilisation, Users of the genetic resources or associated traditional knowledge are required to declare to the relevant competent authority that they have exercised due diligence and to also provide supporting evidence. Suitable points of declaration identified in the recitals to the Regulation include the point when research funds are received by Users and at the final stage of development (which the Regulation envisages to be before requesting marketing approval) of the product that is developed via utilisation of a genetic resource or associated traditional knowledge.

The final stage of development of a product may differ substantially depending on the market sector and, there is room for the Commission to use its implementing powers, as and when needed, to identify a final stage of development in a particular sector.

In addition to the absolute declarations on the part of Users, the appointed competent authority in a member state is also able to carry out checks (including on the spot checks) to verify Users' compliance. The frequency of such checks is presently unclear, but member states are required to ensure that the checks are proportionate and dissuasive.

Enforcement

Users who fail to comply with the requirements imposed by the Regulation will face penalties which shall be set by the relevant member states. Member states are, therefore, required to notify the Commission of their rules on penalties by 11 June, 2015. The Regulation envisages that these penalties should be effective, proportionate and dissuasive.

Comment

Companies working with nature-based ingredients for food, pharmaceutical and personal care products are notable Users under this Regulation, along with universities and academics that use genetic resources for research and development. Any research conducted with respect to identifying beneficial properties of biochemicals extracted from plants, will, for example, require the prior informed consent of the supplier of such plants and a clear mechanism for sharing the benefits of this research with the supplier. Although most companies working with nature-based ingredients already have ethical sourcing practices in place, this would be a good time for such companies to review their relevant practices and policies.

Unsurprisingly, the Regulation is not without its controversy and has been criticised as being excessively burdensome and bureaucratic. The German Plant Breeders' Association, for example, has sought annulment of the Regulation before the Court of Justice of the EU and similar proceedings have also been initiated by Dutch interest groups. The main objection raised by different organisations is that, in practice, compliance with the Regulation's due diligence requirements is onerous as a company that may be working with a significant number of genetic resources at any one time. Keeping a document trail, particularly of every line of plant used in a breeding program, for example, may be practically impossible.

There is some respite; the Regulation will not have retrospective effect and the provisions of the Regulation dealing with the obligations of Users (Article 4), the monitoring of Users' compliance (Article 7) and checks on Users' compliance will not take effect until one year after the date of entry of the Protocol in the EU. Additionally, the Regulation attempts to alleviate administrative and compliance requirements by providing a framework for the establishment of a commission-run, internet-based register of collections of genetic resources. A register would arguably lower the risk of the supply or use of genetic resources where evidence of legal access is inconclusive or lacking. According to the recitals to the Regulation, Users that obtain a genetic resource from a collection included in the register will be considered to have exercised sufficient due diligence. Such collection of genetic resources will be particularly useful for those conducting nature-based research on a smaller scale such as academics, universities, non-commercial researchers, and small and medium-sized enterprises.

Given its vast scope and far-reaching impact, we expect a fair amount of interest and commentary on the Regulation to be put forward by Users and industry organisations across Europe over the next twelve months.

https://www.bristows.com/news-and-publications/articles/protecting-biological-resources-implementation-of-the-nagoya-protocol-in-europe

Vocabulary

recognition 英 [ˌrekəgˈnɪʃn] 美 [ˌrɛkəgˈnɪʃən] n. 认识；识别；承认
rationale 英 [ˌræʃəˈnɑːl] 美 [ˌræʃəˈnæl] n. 理论的说明；基本原理；基础理论
pandemic 英 [pænˈdemɪk] 美 [pænˈdɛmɪk] adj. （疾病）大流行的；普遍的

regulatory requirements 法规要求；管理要求；监管要求；规范要求
diligence obligation 勤勉义务
certificate of compliance 守法证
amongst 英 [ə'mʌŋst] 美 [ə'mʌŋst] prep. (表示位置)处在……中；(表示范围) 在……之内
compliance 英 [kəm'plaɪəns] 美 [kəm'plaɪəns] n. 服从；听从；承诺
envisage 英 [ɪn'vɪzɪdʒ] 美 [ɛn'vɪzɪdʒ] vt. 想象；设想；观察；展望
the absolute declarations on the part of the user 用户部分的绝对声明
proportionate 英 [prə'pɔːʃənət] 美 [prə'pɔːrʃənət] adj. 相称的；适当的
dissuasive 英 [dɪ'sweɪsɪv] 美 [dɪ'sweɪsɪv] adj. 劝戒的
enforcement 英 [ɪn'fɔːsmənt] 美 [ɪn'fɔːrsmənt] n. 强制；实施；执行
burdensome 英 ['bɜːdnsəm] 美 ['bɜːrdnsəm] adj. 累赘的；难以承担的
bureaucratic 英 [ˌbjʊərə'krætɪk] 美 [ˌbjʊrə'krætɪk] adj. 官僚的；官僚作风的
annulment 英 [ə'nʌlmənt] 美 [ə'nʌlmənt] n. 废除；(法院对婚姻等)判决无效
proceeding 英 [prə'siːdɪŋ] 美 [pro'sidɪŋ prə-] n. 诉讼；行动；会议记录
respite 英 ['respaɪt] 美 ['respɪt] n. 休息期间；缓解；暂缓；延期

Useful Expressions

1. comply 英 [kəm'plaɪ] 美 [kəm'plaɪ] vi. 遵从；依从；顺从；应允；同意
The commander said that the army would comply with the ceasefire.
指挥官说军队会遵守停火协议。
2. subsequent 英 ['sʌbsɪkwənt] 美 ['sʌbsɪˌkwɛnt, -kwənt] adj. 随后的；后来的；作为结果而发生的
Those concerns were overshadowed by subsequent events.
随后发生的事使之前所关注的那些问题显得无足轻重。
3. notify 英 ['nəʊtɪfaɪ] 美 ['noʊtɪfaɪ] vt. 通知；布告
The skipper notified the coastguard of the tragedy.
船长向海岸警卫队报告了这起灾难。

Questions

1. Why did the signatories adopt Nagoya Protocol?
2. What is "associated traditional knowledge"?
3. What are the purposes of these various legal instruments?
4. What is required of users according to the Regulation?
5. What must users do without the certification?

Chapter Seven

Regeneration of Medicinal Plant Resources

7.1 Natural Regeneration of Medicinal Plant Resources

Passage 1

Natural Regeneration Status of the Endangered Medicinal Plant, Taxus Baccata Hook. F. syn. T. Wallichiana, in Northwest Himalaya (Part 1)

Introduction

The genus Taxus (Taxaceae) contains seven to nine species, among which, Taxus baccata, T. brevifolia, T. canadensis, T. cuspidata and T. floridana are the most well-known species. T. baccata Hook. F. syn. T. wallichiana Zucc is found in Europe, North Africa, the Caucasus, Iran and temperate areas of the Indian sub-continent. Popularly known as Himalayan yew, it is distributed from Pakistan to southwest China, Nepal and Bhutan, mainly at elevations of 1,800−3,300 m a. s. l. In Indian Himalaya, it is locally called "Rakhal" or "Thuner", and is the only species of the genus in Jammu and Kashmir, Himachal Pradesh, Uttarakhand, Meghalaya, Nagaland and Manipur, mainly associated with oak (kharshu) and silver fir species, rarely with spruce, deodar cedar and oak (mohru) in the western Himalaya, and in eastern Himalaya, and it is mostly associated with silver fir, hemlock-spruce and rhododendron. This evergreen tree flowers during March-April and seeds ripen from September to November. It can tolerate frost, drought and strong winds, and usually grows on limestone-derived moist soil.

Yew is a valuable medicinal plant, chiefly for its Taxine and DAB (10-deacetyl baccatin III) content. Initially, the tree was exploited for bark but later demand shifted to leaves. It is a source of taxine, a precursor to the drug Taxol, used in preparation of Paclitaxel for treatment of ovarian cancer. Oncologists working with taxol regard it as one of the best anticancer agents. According to the Wealth of India, Taxol was first extracted from bark of T. brevifolia, and anti-cancer properties were first reported in 1964. As well as Taxol, the Himalayan yew is a source of the drug Zarnab, a popular medicine of Unani system of medicine. Bark and leaves are used for bronchitis, asthma and insect bites and as an aphrodisiac. In high-altitude Himalayan regions, local inhabitants use the bark for making a traditional tea for treatment of coughs and colds; bark paste is applied externally to cure headache. Yew wood is an extremely hard and durable product used for making furniture, carving and as fuel.

The Himalayan yew has poor regeneration, primarily due to low seed production and poor germination. Other factors like weed growth, accumulation of debris, continuous grazing and infrequent mast years also adversely affect its natural regeneration. Furthermore, the slow initial growth hampers establishment and development in the natural habitats, particularly in fir-kharshu oak forests, especially on cooler aspects. The young plants require shelter and deep shade and, consequently, do not thrive in areas where forests have been cleared. Yew trees can live to a great age because of their slow growth.

To produce one dose of Paclitaxel, the bark from six mature trees is needed. Due to high demand, there is much biotic pressure on yew from lopping, peeling bark and grazing. The extent of canopy damage is likely to have serious consequences for biomass yield, plant survival and natural regeneration. Generally, regeneration of this species is better in moist and shady micro-sites at undisturbed locations. These biotic pressures, along with poor regeneration, are collectively responsible for the rapid decline in populations in the Himalaya have declared this species as "endangered" in the Himalayan region. However, there is a paucity of information on natural regeneration in the Himalaya. This lack of information restricts forest managers and policy makers in formulating conservation and management strategies for this high-value ecologically important but endangered species. Therefore, the present study was undertaken to assess natural regeneration of the Himalayan yew to initiate effective steps for its conservation and management.

Study Area

The study area is in the district Shimla, Himachal Pradesh, a hilly state in north India (30°22′−33°12′N, 75°47′−79°04′E, 200−7,000 m a.s.l.). Topographically, the state is divided into three zones—the Siwaliks or outer Himalaya, the low mountains in comparison to great mountains in the middle, and Zanskar (high peaks). The most important rivers are Chenab (Chandrabhaga), Ravi (Iravati), Sutlej (Shatadru), Beas (Vipasa) and Yamuna (Jamuna). The climate varies from sub-tropical at lower elevations to temperate in mid-elevations and alpine at higher elevations. Rainfall varies from 500 mm in Lahaul and Spiti to over 3,400 mm in Kangra. According to Champion and Seth, the forest types of the state are 12 C1a—lower western Himalayan temperate forest, 12 C1b—Moru oak forests, 12 C1d—western mixed coniferous forests, 12 C2a—Kharshu oak forests, 12 C2b—upper oak and fir forests. All these geographical, edaphic and climatic features support a rich and unique biological wealth in the state, with approximately 3,500 species of flowering plants reported.

Display Full Size

Broadly, the selected study sites are in the northwest ranges of the Himalaya, between 31°03′N and 77°07′E, at an average altitude of 2,800 m. The main forest type of the region is moist temperate, which covers an area of 5,131 km². The climate is predominantly cold during winter and pleasantly warm during summer, with temperature ranging from 4℃ to 33℃. The average temperature during summer is 14℃ to 20℃, and between −7℃ and 10℃ in winter. Rainfall varies between 24.0 mm and 1,020.0 mm, average 900 mm, and average snowfall is around 115 cm.

The study was conducted in three forest divisions at elevations from 2,400 m to 2,800 m. According to Survey of India Toposheets 53E 11/12, 53 E/12 and 52E-8/12, these are (i) Kotgarh forest division (31°0′−31°0′N, 77°29′−77°31′E); (ii) Theog forest division (31°6′−31°7′N, 77°36′−77°37′E) and (iii) Chopal forest division (31°14′−31°15′N, 77°30′−77°33′E).

https://www.tandfonline.com/doi/full/10.1080/21513732.2010.527302?scroll=top&needAccess=true

Vocabulary

taxus 英 [ˈtæksəs] 美 [ˈtæksəs] *n.* 紫杉
Himalayan yew 喜马拉雅紫杉
spruce 英 [spruːs] 美 [sprus] *n.* 云杉；针枞
deodar 英 [ˈdiːədɑː] 美 [ˈdiːədɑː] *n.* 喜马拉雅雪松
cedar 英 [ˈsiːdə(r)] 美 [ˈsidɚ] *n.* 雪松；香柏；香椿；西洋杉
rhododendron 英 [ˌrəʊdəˈdendrən] 美 [ˌroʊdəˈdendrən] *n.* 杜鹃花；[植]杜鹃花属
yew 英 [juː] 美 [joo] *n.* 紫杉；紫杉木
taxine 英 [ˈtæksiːn] 美 [ˈtæksiːn] *n.* 紫杉碱
taxol [医]紫杉酚；红杉醇；红豆杉醇；落羽松醇
ovarian cancer [医]卵巢癌
debris 英 [ˈdebriː] 美 [dəˈbriː] *n.* 碎片；残骸；残渣
mast 英 [mɑːst] 美 [mæst] *n.* 桅杆；橡树果实
paclitaxel 英 [ˌpæklɪˈtæksəl] 美 [ˌpæklɪˈtæksəl] *n.* 特素
paucity 英 [ˈpɔːsəti] 美 [ˈpɔsɪti] *n.* 少量；缺乏；不足
floristic 英 [flɒˈrɪstɪk] 美 [floʊˈrɪstɪk] *adj.* 花的；植物的；植物种类的

Useful Expressions

1. oncologist 英 [ɒŋˈkɒlədʒɪst] 美 [ɒŋˈkɒlədʒɪst] *n.* 肿瘤学家
That's why I'm talking to an oncologist.
这就是我和肿瘤专家谈话的原因。
2. graze 英 [greɪz] 美 [grez] *vt.* 放牧；(让动物)吃草；轻擦；擦破

I had grazed my knees a little.

我的膝盖擦破了一点皮。

3. adversely 英 [əd'vɜːslɪ] 美 [əd'vɜːslɪ] *adv.* 逆地；反对地；不利地；有害地

Nicotine adversely affects the functioning of the heart and arteries.

尼古丁会对心脏及动脉功能造成损害。

Questions

1. What is Himalayan yew?
2. What is yew used for?
3. Where is the useful content stored in yew?
4. How many mature trees are needed to extract one dose of Paclitaxel?
5. Where was the study carried out?

Passage 2

Natural Regeneration Status of the Endangered Medicinal Plant, Taxus Baccata Hook. F. Syn. T. Wallichiana, in Northwest Himalaya (Part 2)

Methods

Sites were mainly selected to contain T. baccata. The study was carried out at two elevations, 2,400−2,600 m and 2,600−2,800 m a. s. l. Within each site, a representative sample plot of approximately 1 ha was selected and sample plots of 20 m×20 m were laid containing 16 quadrats of 2 m×2 m per plot to count recruits (r), defined as current-year seedlings, and unestablished regeneration (u), seedling other than recruits that has not established and were less than 2m high. Four unestablished plants were taken as equivalent to one established plant, and established plants (e) were those more than 2m high. The height of unestablished plants was measured for assessment of regeneration. Regeneration data for T. baccata and associated species were collected on the basis of seedlings, saplings and pole stage trees in each quadrat. The criteria followed for regeneration assessment are given.

Regeneration Assessment

The data were analysed in each sample plot for assessment of regeneration status of yew and associate species. Establishment of plants at 2,500 per hectare was considered a satisfactory level of regeneration. A quadrat was considered fully stocked when it contained one established plant. The data thus collected were analysed using the formulae of Chacko, below.

Weighted average height (m)=

$$\frac{\text{Total height of unestablished regeneration}+(\text{No. of established plants}\times\text{Establishment height})}{\text{Total unestablished plants}+\text{Total established plants}}$$

$$\text{Recruits }(r)\text{ per hectare}=2500\sum_{i=1}^{n}\frac{ri}{m}$$

$$\text{Unestablished regeneration }(u)\text{ per hectare}=2500\sum_{i=1}^{n}\frac{ui}{m}$$

$$\text{Established regeneration }(e)\text{ per hectare}=2500\sum_{i=1}^{n}\frac{ei}{m}$$

In the formulae, N is the number of sampling units, m the total number of recording units in survey, ri the total number of recruits in each sampling unit, ui the total number of unestablished plants in each sampling unit and ei the total number of established plants in each sampling unit.

$$\text{Establishment index }(I_1)=\frac{\text{Weighted average height}}{\text{Establishment height}}$$

From the about estimates, the following indices were calulated:

$$\text{Sleeking index }(I_2)=\frac{1}{2500}\times\frac{\text{Unestablished regeneration (per ha)}}{4}+\text{Established regenerarion (per ha)}$$

Established stocking percent$=100(I_1\times I_2)$

Site Assessment

The floristic composition was studied in sample plots of 20 m × 20 m (four plots per elevation). The phytosociological parameters of relative density, relative frequency, relative basal area and girth/diameter were recorded for different plant communities using different formulae of Raunkiaer.

The importance value index (IVI), which is an integrated measure of relative frequency, relative density and relative basal area [IVI = relative basal area (RBA) + relative density (RD) + relative frequency (RF)] was obtained from percentage frequency, density and basal area according to Phillips. Although, the IVI was calculated for both tree and shrub species.

Results and Discussion

Assessment of natural regeneration is an important tool to evaluate stocking, competition problems and composition of new stands under forest management. In the present study, mean natural regeneration status of T. baccata and associated species was assessed in three forest divisions in Himachal Pradesh, India. In all forest divisions, Abies pindrow, Picea smithiana, Pinus wallichiana, Quercus dilatata, Q. semecarpifolia and Juglans regia were associated with T. baccata. A total of eight shrub species, Rhamnus triquetrus, Rosa macrophylla, Berberis aristata, Sarcococca saligna, Viburnum cotinifolium, Cotoneaster bacillaris, Prinsepia utilis and Daphne cannabina, were present in the sites. In general, at all the elevations and forest divisions, V. cotinifolium and D. cannabina were dominant shrub species.

Both upper (2,600－2,800 m) and lower (2,400－2,600 m) elevations of Kotgarh forest division had most recruits (527/ha and 417/ha, respectively), followed by Theog and Chopal forest divisions. Individually, P. smithiana had most recruits at both upper and lower elevations (898/ha and 800/ha) in Kotgarh forest division, followed by A. pindrow (781/ha and 644/ha, respectively) and T. baccata (781/ha and 488/ha, respectively). As in Kotgarh forest division, the number of recruits of T. baccata in Theog forest division (429/ha and 390/ha) followed that of P. smithiana and A. pindrow and in Chopal forest division (195/ha and 136/ha), then P. wallichiana and Q. dilatata. The lower number of recruits of T. baccata at lower elevation may be due to a higher density of broadleaf species and high biotic influence in the area. This may also be due to low crown competition in lower elevations. The maximum mean height (63.22 cm) and establishment index (0.314) was recorded at the upper elevation in Chopal forest division, with most A. pindrow (113.85 and 0.565, respectively). In all three forest divisions, only recruits of T. baccata were found and there was no established regeneration of this species. This may be due to over-exploitation for pharmaceutical purposes by locals and trampling by grazing animals, as also suggested by Shamet and Gupta. They investigated natural regeneration status of T. baccata in different forest areas, in which established natural regeneration was very poor in Shimla forest, only 250 plants/ha in Chanjar forest and no regeneration in Baggi and Tikker forests. Similarly, Kopp investigated seedling performance for 5 years and reported that without fencing and opening up the canopy (adequate light), vigorous natural regeneration is not possible.

https://www.tandfonline.com/doi/full/10.1080/21513732.2010.527302?scroll=top&needAccess=true

Vocabulary

current-year seedling 当年苗
quadrat 英 [ˈkwɒdrət] 美 [ˈkwɒdrət] n. 嵌块（填空的铅块）
formulae 英 [ˈfɔːmjəliː] 美 [ˈfɔːmjəliː] n. 配方；公式(formula 的名词复数)；分子式；方案
phytosociological parameter 植物群落学参数
basal area 底面积；断面积
girth 英 [gɜːθ] 美 [gɜːrθ] n. 周长
statistical plant geography 统计植物地理学
importance value index 重要值指数
integrated measure of relative frequency 相对频率综合测量

crown width　顶宽
lux meter　照度计
canopy 英 [ˈkænəpi] 美 [ˈkænəpi] n. 天篷；华盖；苍穹；树荫
grazing animal　草食动物；放牧家畜

Useful Expressions

1. recruit 英 [rɪˈkruːt] 美 [rɪˈkrut] vt. 招聘；征募；雇用
The police are trying to recruit more black and Asian officers.
警方正在试图招募更多黑人和亚裔警官。
2. trample 英 [ˈtræmpl] 美 [ˈtræmpəl] vi. 践踏；重重地踩；脚步沉重地走
They say loggers are destroying rain forests and trampling on the rights of natives.
他们说伐木工人正在毁坏雨林,践踏土著居民的权利。

Questions

1. What is the principle of site selection?
2. What is the satisfactory level of regeneration?
3. What is the significance of assessing natural regeneration?

7.2　Artificial Regeneration of Medicinal Plant Resources

Passage 1

in vitro Regeneration of Hemidesmus Indicus L. R. Br, an Important Endangered Medicinal Plant (Extract)

Introduction

The maximum mean stocking index (0.08) and mean established stocking percentage (2.20%) were recorded at the lower elevation in Chopal and Kotgarh forest divisions, respectively. This may be due to the large organic matter layer and crown projection ratio at higher elevation in these forest areas. Jha et al. noted that a thick layer of humus affected natural regeneration process of fir and spruce in Narkanda forests of Himachal Pradesh.

The IVI of T. baccata increased from lower to higher elevations, i.e. 86.5−92.2 in Kotgarh forest division and 40.2−70.9 in Theog forest division. Similar results were found for the associated species of T. baccata in Theog and Chopal forest divisions. The results were in line with Gupta for the IVI value of A. pindrow, which increased with elevation in Narkanda and Chhachpur forest areas. In contrast, in Kotgarh forest division, the IVI value of associated species of T. baccata decreased from lower (IVI 71.1) to higher (IVI 66.8) elevation. The decreasing of IVI with increasing elevation was due to larger trees at lower elevation as well as the presence of more tree species. Similarly, the established stocking percentage increased from lower to higher elevation in Kotgarh forest division and decreased from lower to higher elevation in Theog and Chopal forest divisions. This may be due to light availability within these areas.

The soil at lower elevations had a high organic carbon, available nitrogen and available potassium content, in addition to good moisture and pH levels. This improved soil may be related to the presence of several broadleaf species with fast leaf decomposition, which boosted seedling regeneration beneath the canopy. In general there was a decrease in availability of moisture, organic carbon (approximately 40%) available nitrogen (approximately 25%) and available phosphorus (approximately 9%) with the increase in soil depth, whereas available potassium (approximately 2.5%) and soil pH increased with the increase in soil depth at all sites.

It was generally found that anthropogenic pressure on the lower elevation sites was very high compared to that on higher elevation sites. In all forest divisions at lower elevations, ruthless harvesting through cutting of twigs and branches, bark peeling and, in some places, cutting of the whole tree clearly indicated that the species was facing a high risk of extinction. Besides grazing pressure, human activities such as camping and deforestation for timber and road construction were very high at lower elevations. Based on the above observations, it can be concluded that the lower elevations, in spite of being conducive for T. baccata growth, are under various biotic pressures and thus unable to support optimum regeneration of yew.

In the Mediterranean mountains of southern Spain, T. baccata is catalogued as an endangered species prone to extinction due to the small size and senescent status of most populations. This study investigated the effects of herbivory (by wild goats and domestic livestock) and the protective role of woody shrubs on regeneration of yew in an autochthonous montana pine (Pinus sylvestris var. nevadensis) stood with a dense shrub understorey. The estimated density of yew in the study plot was 288 individuals/ha, more than 90% as juveniles (seedlings and saplings) which were mostly under fleshy-fruited shrubs. Saplings suffered serious herbivore damage when unprotected by shrubs. Thus, these shrubs were the best habitat for yew seedling establishment and sapling survival and growth. Maintenance of healthy populations of yew in the Mediterranean mountains was strongly dependent on the conservation of a well developed fleshy-fruited understorey and the associated avian dispersers.

In the Himalayan, there was much biotic pressure from lopping trees in forests by the local population, and also from their large herds of sheep, goats and cattle. The movement of herds and lopping of trees were common, resulting in poor growth and confined populations of T. baccata in the region, while studying population ecology of yew in

Khokhan Wildlife Sanctuary, Himachal Pradesh, found that 50% of surviving yew trees had been affected by bark removal and the remaining trees were subject to lopping and felling for fuel. Rikhari et al. studied stand structure, canopy density, regeneration and suggested conservation strategy for T. baccata in the western part of Uttarakhand Himalaya. They found that recent uncontrolled harvesting from the wild for the extraction of taxol was wiping out this species since it was very slow growing and had poor regeneration. The extent of canopy damage was likely to have serious consequences for biomass yield, plant survival and natural regeneration. The regeneration of this species was better in moist and shady microsites in undisturbed locations. Unfortunately, Himalayan yew had very thin bark as compared to some other yew species, making it also susceptible to fire.

Based on the above data, it can be concluded that this yew species is facing biotic pressure and therefore it is important to develop location-specific strategies and action plans for successful regeneration, vigorous stand growth, sustainable utilisation and thus, conservation of T. baccata.

Conclusions

The regeneration of T. baccata was very poor in all three forest divisions at 2,600 m to 2,800 m a. s. l. in northwest Himalaya. Over-exploitation of bark and leaves for pharmaceuticals by locals and trampling by grazing animals were important reasons for poor regeneration, particularly at lower elevations. Adequate recruitment and regeneration of associated species imply that yew is being over-exploited. Overall, these observations indicate that natural populations of T. baccata, particularly at lower elevations, are under severe threat of over-harvesting and other anthropogenic pressures, leading to poor regeneration. There is an immediate need to protect and manage Taxus forests from lopping, peeling bark, grazing and other destructive activities. Besides in-situ conservation and management, mass afforestation with this species in protected forest areas, particularly at religious groves where harvesting of green trees is generally not allowed, need to be undertaken with the participation of local communities. In view of poor regeneration from seed, propagation protocols should be extensively used for artificial regeneration (through seeds and vegetative parts) to develop mass planting material. Effective execution of a location-specific strategy and action plans with participation of local communities would be useful for conservation management and sustainable utilisation of this species.

http://www.ukjpb.com/article_details.php?id=53

Vocabulary

humus 英 ['hju:məs] 美 ['hjuməs] *n.* 腐殖质
avian disperser 鸟类分散剂

felling 英 [ˈfelɪŋ] 美 [ˈfelɪŋ] n. 咬口折缝；二重接缝

Useful Expressions

1. in line with   本着；跟……一致；符合

The structure of our schools is now broadly in line with the major countries of the world.

我们学校的建构现在基本上与世界主要国家的相一致了。

2. removal 英 [rɪˈmuːvl] 美 [rɪˈmuvəl] n. 除去；搬迁；免职；移走

What they expected to be the removal of a small lump turned out to be major surgery.

他们原以为只需切除小肿块的手术最后变成了一次大手术。

3. be subject to   v. 受支配；从属于；可以……的；常遭受……

Employee appointment to the Council will be subject to a term of probation of 6 months.

被任命到理事会的员工将有6个月的见习期。

Questions

1. What is the deference between anthropogenic pressure on the lower elevation sites and that in higher elevation sites?

2. What causes biotic pressure in the Himalayan?

3. What should be done for conservation management and sustainable utilisation of T. baccata?

Passage 2

An Improved Protocol for *in vitro* Propagation of the Medicinal Plant Mimosa Pudica L. (Extract)

Materials and Methods

Plant material and establishment of *in vitro* micropropagation procedure:

in vitro establishment of aseptic cultures of M. pudica was carried out by seeds harvested from dry and mature fruits, from plants developed in the natural environment. After collection, the seeds were subjected to surface asepsis by rinsing in tap water for 60 minutes. Then, the seeds were immersed in ethanol 70% (v/v) for 30 seconds and, later, in a bleach solution with 2% of active chlorine, diluted at 30% (v/v) for 15 minutes. Finally, the seeds were washed in distilled and autoclaved water. After asepsis, the seeds were inoculated in MS basal medium, without growth regulators, in a laminar flow hood. The success of *in vitro* establishment was estimated 30 days after inoculation, considering the percentages of seeds contamination and total germination. In order to obtain stabilized cultures, 60 days after *in vitro* establishment, nodal segments (2—3 cm) from seedlings were transferred to MS culture medium supplemented with α-naphthalene acetic acid (NAA: 0.107 μM) or 6-benzylaminopurine (BAP: 2.22 μM), besides the control group (without growth regulators). After 60 days, the *in vitro* cultures were evaluated regarding the number of shoots per explants, height of shoots, and percentage of rooted plantlets.

in vitro Multiplication

Aiming to eliminate the residual effects of previous culture media, after *in vitro* stabilization phase, the explants were transferred and kept for 30 days on MS medium without growth regulators. Then, nodal segments (2—3 cm) were obtained from plantlets maintained in this condition and transferred to MS culture medium supplemented with BAP, kinetin (KIN), adenine sulphate (AS), diphenylurea (DFU) or thidiazuron (TDZ), at 0, 2.5, 5 or 7.5 μM, totaling 16 treatments. The cultures were kept in these culture media for 30 days and evaluated regarding the number of shoots and roots per explant, height of shoots, and callus development. Root quality was also evaluated with scores, ranging from 1 to 5 given by three independent evaluators, with 5 corresponding to the root system that presented the best development, and 1 for explants that did not present root development.

After we found that BAP was the cytokinin that provided the highest *in vitro* multiplication rates, new assays were carried out aiming to further increase culture proliferation. Nodal segments (2—3 cm) obtained from explants aseptically established in MS medium without growth regulators, were inoculated in MS medium supplemented with BAP (0, 5 or 7.5 μM) and NAA (0, 0.05, 0.25 or 0.5 μM), in all possible combinations, totaling 12 treatments. The cultures were evaluated after 45 days regarding percentage of explants presenting three or more shoots, number and height of shoots.

Aiming to obtain elongated shoots, nodal segments (2—3 cm) from the best treatment in the previous assay (5 μM BAP plus 0.5 μM NAA) were inoculated in MS medium supplemented with gibberellic acid (GA3: 0, 0.28, 1.44, 2.89 or 4.53 μM), in presence or absence of 5 μM BAP, totaling 10 treatments. After 45 days, the cultures were evaluated regarding number and height of shoots and the number of roots per explant.

Considering the results found in the assay with the different cytokinins, another set of explants, previously kept in MS medium without growth regulators, was transferred to MS medium supplemented with TDZ (0.6, 0.9 or 1.2 μM) plus NAA (0, 0.05, 0.25 or 0.5 μM), in all possible combinations, totaling 12 treatments. After 45 days, the cultures were evaluated considering the number of shoots higher than 0.5 cm. As the new shoots, in this assay, did not show suitable elongation, the explant clusters (rosettes) produced (0.5 ± 0.1 cm) were transferred from MS culture media with 0.6 μM TDZ, singly or combined with NAA, to MS culture media without growth regulators or to MS culture media supplemented with 0.25 μM NAA. After 45 days, the cultures were evaluated considering the percentage of regenerated shoots and height of shoots.

in vitro Rooting

Aiming to stimulate the rooting of micro-cuttings, nodal segments from plantlets (2—3 cm, excluded the apical part) were transferred to MS culture media supplemented with NAA, indole-3-acetic acid (IAA) or indole-3-butyric acid (IBA) (0.1, 0.2, 0.3 or 0.4 μM, besides the control). After 40 and 60 days of inoculation, the cultures were evaluated regarding the number of roots and the length of the largest root.

in vitro Culture Conditions

At all stages of *in vitro* culture, the plantlets were kept in 2.5 cm×15 cm test tubes. The MS culture media was supplemented with MS vitamins, sucrose (30 g L1), and agar (7 g L1). The culture media pH was adjusted to 5.7 ± 0.1 before autoclaving, carried out for 20 minutes at 120℃ and 1 atm of pressure. The test tubes were capped with autoclaving polyethylene closures, and sealed with PVC film (Vitaspenser, Goodyear, 15 μm). The cultures were kept in a growth room under controlled conditions of temperature (26 ± 1℃), photoperiod (16 h) and luminosity (40 μmols m-2 s-1).

Ex Vitro Acclimatization

After 60 days of cultivation in rooting medium, the plantlets were removed from test tubes and their roots were washed in running tap water to remove culture media debris. Later, the plantlets were transplanted to polystyrene trays with 128 cells, filled with commercial substrate Plantmax Hortalicas. The trays were covered and wrapped with transparent plastic, remaining for 20 days in shadowed environment. After this period, the trays were transferred to a greenhouse covered with transparent plastic and Sombrite 70%, and maintained under programmed micro sprinkler system undertaken for 5 minutes twice a day. After 40 days, the plants were transferred to pots with a mixture of soil/washed sand/cattle manure at the proportion of 3:2:1 (v/v/v). The efficiency of acclimatization procedures was evaluated taking in account the plantlets final survival percentage.

Statistical Analysis

All experiments were conducted in a completely randomized design, with five replicates, except for cases that percentages were compared, in which ten repetitions per treatment were used. Linear or polynomial regression in accordance with residual requirements in assays related to multiplications and rooting phases was employed. In some of the experiments, counting data were normalized by the equation and the results submitted to analysis of variance (ANOVA). The obtained means were compared through Scott-Knott test at 5% probability, using SAEG software (version 9.1).

http://www.academicjournals.org/journal/AJB/article-full-text/EC56A6863009

English Reading of Pharmaceutical Botany Resource

Vocabulary

micropropogation 英 [maɪkrəprɒpəˈgeɪʃn] 美 [maɪkrəprɒpəˈgeɪʃn] n. 微繁殖
aseptic 英 [ˌeɪˈseptɪk] 美 [əˈseptɪk, e-] n. 防腐剂
asepsis 英 [æˈsepsɪs] 美 [æˈsepsɪs] n. 无菌；无病毒
tap water　自来水；非蒸馏水
ethanol 英 [ˈeθənɒl] 美 [ˈeθənoʊl] n. 乙醇
bleach solution　漂白剂；消毒液
active chlorine　活性氯；有效氯
laminar flow hood　层流净化罩
nodal segment　（茎）节段
residual 英 [rɪˈzɪdjuəl] 美 [rɪˈzɪdʒuəl] adj. 残留的；残余的
proliferation 英 [prəˌlɪfəˈreɪʃn] 美 [prəˌlɪfəˈreɪʃn] n. 增生；增殖；分芽繁殖；再育
gibberellic acid 英 [ˌdʒɪbəˈrelɪk] 美 [ˌdʒɪbəˈrelɪk] n. 赤霉酸（一种植物生长调节剂）
polystyrene 英 [ˌpɒlɪˈstaɪriːn] 美 [ˌpɑːliˈstaɪriːn] n. 聚苯乙烯
replicate 英 [ˈreplɪkeɪt] 美 [ˈreplɪket] n. 复制品

Useful Expressions

1. autoclave 英 [ˈɔːtəʊkleɪv] 美 [ˈɔːtoʊkleɪv] n. 压热器；高压蒸气灭菌器
We also can use this technology for your autoclave plant.
我们还可以利用此技术改造你的高压釜。
2. stabilize 英 [ˈsteɪbəlaɪz] 美 [ˈstebəˌlaɪz] vi. 变得稳定（稳固或固定）
Although her illness is serious, her condition is beginning to stabilize.
虽然她病得很重，但病情正开始趋于稳定。
3. randomize 英 [ˈrændəmaɪz] 美 [ˈrændəˌmaɪz] v. 使随机化；（使）任意排列
Properly randomized studies are only now being completed.
真正的随机研究现在才接近完成。

Questions

1. How was root quality evaluated?
2. When were the plantlets transplanted to polystyrene trays?
3. What was SAEG software used for?

7.3 Organ Regeneration of Medicinal Plant Resources

Passage 1

What Processes Contribute to Organ Regeneration?

The processes that contribute to organ regeneration involve stimulating growth in specific cells. By studying the regenerative abilities of certain animals, researchers obtain a better understanding of what human bodies require to repair or regrow tissue. Regenerative medicine research has revealed the role of cytokines, growth factors, stem cells, and other factors that play in tissue regeneration.

Since the 18th century, scientists have marveled at the unusual ability of certain animals to undergo cell regeneration. Newts can regrow a severed limb, and salamanders can replace a missing tail. Some species of fish can regrow a damaged fin. A new planaria worm will grow from each piece of a dissected planaria worm.

Three factors contribute to organ regeneration in animals. Organ and other tissue cells that normally do not reproduce will do so during injury or illness. In certain animals, cells transform from one type of tissue into another. Stem cells are also involved in these regenerative processes. In comparison, human bodies have the capacity to heal, though not without scarring.

Once scarring occurs, cell growth generally ceases. Researchers discovered a means of inhibiting this process by developing a substance called extracellular matrix that contains connective tissue, pig bladder cells, and proteins. When applied to the severed finger of a patient, the substance prevented scarring and triggered reproduction of various types of cells in the digit. In about four weeks, the patient reportedly regrew the entire portion of the severed finger.

Organ regeneration usually begins with placing specific tissue cells, along with a growth medium, into petri dishes. Tissue that develops from the cells is placed over a specially designed foundation. Depending on the type of regenerated tissue, mature growth usually occurs in about eight weeks. Surgeons typically implant the entire specimen into the patient's body, including the foundation scaffold. The scaffold generally dissolves and the new tissue functions without the possibility of rejection.

Using this technique for tissue regeneration, physicians have successfully grown dozens of animal and human tissue types. Blood vessels, connective and muscle tissue, as well as bladders, are among the regenerative advancements that patients receive surgically. These laboratory grown replacements all began as cells donated by the patients themselves.

In addition to the benefit of not having to take antirejection medications, natural organ regeneration increases the number of options for patients on transplant lists. The number of patients requiring donated organs generally exceeds the number of available organs. Advancements in regenerative medicine will allow patients the opportunity to grow their own body parts.

http://www.wisegeek.com/what-processes-contribute-to-organ-regeneration.htm

Vocabulary

cytokine 英 [ˌsɪtəˈkɪn] 美 [ˌsɪtəˈkɪn] *n.* 细胞活素
severed limb　断肢
salamander 英 [ˈsæləmændə(r)] 美 [ˈsæləˌmændə] *n.* 火蜥蜴；火怪；耐火的人
fin 英 [fɪn] 美 [fɪn] *n.* 鱼鳍；散热片；鳍状物
planaria 英 [pləˈneərɪə] 美 [pləˈneərɪr] *n.* 涡虫；真涡虫
extracellular matrix　细胞外基质
pig bladder cell　猪膀胱细胞
bladder 英 [ˈblædə(r)] 美 [ˈblædər] *n.* 膀胱

Useful Expressions

1. regenerative 英 [rɪˈdʒenərətɪv] 美 [rɪˈdʒenəˌretɪv, -ərətɪv] *adj.* 再生的；恢复的；新生的

How these biological findings may inform regenerative medicine and stem-cell biology and therapeutics will be considered and discussed.

对于这些生物的研究结果会如何告知再生医学、干细胞生物学和疗法，将被审议和讨论。

2. scaffold 英 [ˈskæfəʊld] 美 [ˈskæfoʊld] *n.* 脚手架；[史]断头台

Moore ascended the scaffold and addressed the executioner.

穆尔走上断头台，和刽子手说话。

Questions

1. What are the three factors that contribute to organ regeneration in animals?
2. What happens when scarring occurs?
3. How did physicians manage to grown dozens of animal and human tissue types?

Passage 2

Rapid *In Vitro* Plant Regeneration from Leaf Explants of Launaea Sarmentosa (Willd.) Sch. Bip. ex Kuntze

Introduction

Launaea sarmentosa (Willd.) Sch. Bip. ex Kuntze of Asteraceae has been traditionally used as a folk remedy in India. The roots are exploited for the cure against jaundice, as a lactagogue and to purify blood, especially by the tribes of the Western Ghats, a biodiversity hotspot. It has also been used as a coolant, diuretic and demulcent against allergic infections. The habitat of the herb is under increasing anthropogenic, as well as natural pressures, and hence calls for proper management through local measures of protection.

The development of *in vitro* techniques for the culture of isolated plant organs, tissues and cells have led to several exciting opportunities, as well as applications in plant biotechnology. They have allowed widespread use of cell cultures for genetic improvement and plant propagation for production of commercially useful products. Plant propagation via organogenesis is a relatively recent technique used to mass-produce plantlets within a short span of time. The development of a protocol for direct shoot organogenesis is not only useful in micro-propagation, but also opens up alternative ways of introducing novel traits via genetic transformation. Direct organogenesis of various plant species including many medicinal plants has been well reported. The present investigation was carried out to standardize a rapidly reproducible protocol that can be employed at a commercial scale for clonal propagation. Moreover, it also aims at determining the most appropriate growth regulator concentration and combination that can induce direct regeneration of leaf midrib as a source of explant.

Materials and Methods

L. sarmentosa was collected from the coastal area of South Arcot District, India by Mahesh and grown in the experimental garden of the Department of Plant Biology and Plant Biotechnology, St Joseph's College, Tiruchirappalli, India. Three-day-old (0.5−1 cm diameter) leaf midribs were harvested and washed thoroughly with running tap water for 15 minutes before being washed with aqueous solution of 5% (v/v) Teepol (Reckitt Benckiser, India) for 3 minutes. Then they were washed with 1% (v/v) Bavistin (BASF Ltd., Mumbai) for 3 minutes, followed by rinsing with sterile distilled water and 70% (v/v) ethanol for 1 minute. Finally, the explants were given five washes in sterile distilled water. The explants were disinfected with 0.1N $HgCl_2$ and a final washing was given in sterile distilled water. The midrib explants were dissected out and blotted on sterile filter paper discs before inoculation.

The explants were cultured initially on Murashige and Skoog medium containing 3% sucrose (w/v) and 0.8% agar (w/v). The medium was supplemented with 6-benzylaminopurine (BAP; 0.1—1 mg/l), alone or in combinations with naphthalene acetic acid (NAA; 0.1—0.5 mg/l) at pH 5.7. Culture vials of 150 mm×20 mm size containing 15 ml medium with the explants were maintained in a growth chamber at 25±2℃ with a photoperiod of 16 hrs of 40 μ mol m^{-2}s^{-1} white fluorescent lights (Philips, India). The same conditions were continued for shoot multiplication and rooting experiments later. Subsequently subcultures were performed at three week intervals on fresh replenished medium during the regeneration periods. Multiple shoot induction was initiated on MS medium supplemented with BAP (0.1—1 mg/l) and NAA (0.1—0.5 mg/l). A control treatment without cytokinins was also included. Successful observations were recorded from 4 weeks and the frequency of microshoot proliferation was determined as percentages of responding explants and the number and length of shoots per explant.

Roots were initiated on 3—5 cm elongated individual shoots that were isolated from the subcultures and carefully transferred to MS medium containing indole-3-butyric acid (IBA, 0.1—1 mg/l). After 4 weeks, calculations were made on rooted shoots, number and length of roots per shoot. Well developed rooted plantlets were transplanted to plastic pots containing an autoclaved mixture (1:1:1) of garden soil, sand and vermiculite, and kept in a growth chamber under a day-night temperature of 25±2℃ and a photoperiod of 16hrs 40ì mol m^{-2}s^{-1} cool white fluorescent lights. All the plants were irrigated with MS basal salts every three days. The hardened plants were maintained under controlled conditions for 30 days and they were acclimatized in garden. Experiments were set up in a Randomized Block Design (RBD). Each experiment had triplicates and each replicate consisted of at least twenty explants. Ten to fifteen explants were used per treatment in each replication. Mean treatment was compared using Duncan's Multiple Range Test (DMRT) at $P<0.05$ probabilities as per Gomez and Gomez.

Results and Discussion

The responses of leaf explants regarding direct organogenesis and growth are presented in the following. Shoot multiplication was found to be very difficult with leaf explants because the leaves were located in the basal rosette at ground level. The percentage of responding explants during the multiplication phase was high (70%—100%) regardless of the cytokinin (BAP) used and the concentration maintained. Vials without growth regulator produced on an average of 5.8±0.7 shoots per explant after four weeks. In contrast, the presence of BAP resulted in the formation of a larger number of shoot buds. Within two weeks of incubation, direct organogenesis was observed on the surface of leaf explants 0.5 mg/l of BAP resulted in higher number of shoots with an average of 12.6±0.7 and 2.0±0.05 cm length of the shoots per leaf explant. BAP is a very effective growth regulator for shoot multiplication among medicinally important plants such as Pterocarpus marsupium and Watsonia sp. A combined BAP (0.5 mg/l) and NAA (0.2 mg/l) produced 21 ± 0.4 microshoots per explant and 4.2 ± 0.2 cm long shoots. It appears the explants of L. sarmentosa required high cytokinin (0.5mg/l BAP) and low auxin (0.2mg/l NAA) for shoot multiplication. The recorded results are in agreement with previous reports observed with high BAP combined with low NAA for high quality shoots.

Rooting of elongated shoots was observed with IBA (0.1−1 mg/l). On hormone free MS basal medium, the response was low and the mean number of roots was 2.3 ± 0.7 and length of roots was 1.0 ± 0.1 cm per shoot after six weeks of culture. The best rooting response was obtained after four weeks of culture with IBA (0.5 mg/l). The majority of roots developed in 4 weeks. Similar results were reported earlier in Embelia ribes, Penthorum chinense and Sterculia urens. The *in vitro* cultured plantlets that were directly transferred to foam cups showed a high rate of mortality. The potting mix was composed of garden soil: sand: vermiculite (1:1:1) and it was covered with transparent polythene bags for two weeks to ensure higher humidity. The potted plantlets that were kept under *in vitro* conditions were found to be hardened for survival. The procedure employed to acclimatize the transplanted plantlets ended up with successfully hardened plants that were established in garden with a 90% survival rate.

The protocol established in this study can be used for the efficient multiplication of L. sarmentosa. Plantlets produced from this protocol will contribute to the rehabilitation of L. sarmentosa and help to a greater extend to reduce the pressure on the natural population.

https://scielo.conicyt.cl/scielo.php?script=sci_arttext&pid=S0716-97602012000200004&lng=en&nrm=iso&tlng=en

Vocabulary

Launaea sarmentosa　匍枝栓果菊
lactagogue 英 [ˈlæktəgɒg] 美 [ˈlæktəgɒg] *adj.* 催乳的
coolant 英 [ˈkuːlənt] 美 [ˈkulənt] *n.* 冷冻剂；冷却液；散热剂
demulcent 英 [dɪˈmʌlsnt] 美 [dɪˈmʌlsənt] *n.* 镇痛剂；缓和剂
allergic infection　过敏性感染
midrib 英 [ˈmɪdrɪb] 美 [ˈmɪdˌrɪb] *n.* （叶的）中脉
organogenesis 英 [ˌɔːgənəʊˈdʒenɪsɪs] 美 [ˌɔːgənoʊˈdʒenəsɪs] *n.* 器官发生；器官形成

Useful Expressions

1. blot 英 [blɒt] 美 [blɑːt] *vt.* 涂抹；玷污；弄脏；吸掉
Before applying make-up, blot the face with a tissue to remove any excess oils.
上妆前，先用面巾纸把脸上多余的油吸干。

2. standardize 英 [ˈstændədaɪz] 美 [ˈstændərdaɪz] vt. 使标准化；用标准校检

He feels standardized education does not benefit those children who are either below or above average intelligence.

他认为标准化教育对那些智力水平低于或超出平均水平的儿童没有好处。

3. inoculation 英 [ɪˌnɒkjʊˈleɪʃn] 美 [ɪˌnɑkjəˈleʃən] n. ［植］接芽；［医］预防接种

Liquid-quenching apparatus was used to investigate the inoculation process in cast iron.

采用液淬法研究了铸铁的孕育过程。

4. incubation 英 [ˌɪŋkjuˈbeɪʃn] 美 [ˌɪnkjəˈbeʃən, ˌɪŋ-] n. 孵化；(传染病的)潜伏期；［生］(细菌等的)繁殖

Both incubation temperature and substrate humidity affected hatching success and shell crack rate significantly.

孵化温度和湿度显著影响孵化成功率和卵壳龟裂率。

Questions

1. What is the use of Launaea sarmentosa (Willd.) Sch. Bip. ex Kuntze of Asteraceae in India?

2. What is the use of plant propagation via organogenesis?

3. What medium is best for root developing?

Chapter Eight

Development of Medicinal Plant Resources

8.1 Drug Development

Passage 1

Developing Drugs from Traditional Medicinal Plants

Over three quarters of the world's population rely mainly on plants and plant extracts for health care. Approximately one-third of the prescription drugs in the US contain plant components, and more than 120 important prescription drugs are derived from plants. Most of these drugs were developed because of their use in traditional medicine. Economically, this represents $\$8,000-10,000$ M of annual consumer spending. Recent World Health Organization (WHO) studies indicate that over 30 percent of the world's plant species have at one time or another been used for medicinal purposes. Of the 250,000 higher plant species on Earth, more than 80,000 species are medicinal. Although traditional medicine is widespread throughout the world, it is an integral part of each individual culture. Its practice is based mainly on traditional beliefs handed down from generation to generation for hundreds or even thousands of years. Unfortunately, much of this ancient knowledge and many valuable plants are being lost at an alarming rate. The scientific study of traditional medicines and the systematic preservation of medicinal plants are thus of great importance.

For quite a long time, the only way to use plant medicines was either direct application or the use of crude plant extracts. With the development of organic chemistry at the beginning of the 20th century, extraction and fractionation techniques improved significantly. It became possible to isolate and identify many of the active chemicals from plants. In the 1940s, advances in chemical synthesis enabled the synthesis of many plant components and their derivatives. In western countries, it was thought that chemical synthesis of drugs would be more effective and economical than isolation from natural sources. Indeed, this is true in many cases. However, in many other cases, synthetic analogues are not as effective as their natural counterparts. In addition, some synthetic drugs cost many times more than natural ones. Inspired by these realisations, coupled with the fact that many drugs with complex structures may be totally impossible to synthesise, there is now a resurgent trend of returning to natural resources for drug development.

Important Prescription Drugs from Plants

Ephedrine is the oldest and most classic example of a prescription drug developed from a traditional medicinal plant. It is derived from Ma Huang, a leafless shrub. Used to relieve asthma and hay fever in China for over 5,000 years, it was introduced into western medicine in 1924 by Chen Kehong and Schmidt. Ephedrine is an alkaloid closely related to adrenaline, the major product of the adrenal gland. Pharmacologically, Ephedrine is used extensively to stimulate increased activity of the sympathetic nervous system. It is used as a pressor agent to counteract hypotension associated with anaesthesia, and as a nasal decongestant. The drug action of this medicine is based both on its direct effect on α and β adrenergic receptors and on the release of endogenous noradrenaline.

Digitalis is one of the most frequently used medications in the treatment of heart failure and arrhythmia. It increases the contractility of the heart muscle and modifies vascular resistance. It also slows conduction through the atrioventricular node in the heart, making it useful in the treatment of atrial fibrillation and other rapid heart rhythms.

Digitalis is found in the leaves and seeds of Digitalis purpurea and Digitalis lanata, commonly known as the foxglove plant. Foxglove has been used in traditional medicine in many parts of the world—by African natives as arrow poisons, by the ancient Egyptians as heart medicine, and by the Romans as a diuretic, heart tonic, emetic and rat poison. The Chinese, who found this ingredient not only in plants but also in the dry skin and venom of the common toad, has used it for centuries as a cardiac drug. In the western world, the foxglove was first mentioned in 1250 in the writing of a physician, Walsh, and it was described botanically in the 1500s.

Digitalis is a glycoside containing an aglycone, or genin, linked to between one and four sugar molecules. The pharmacological activity resides in the aglycone, whereas the sugar residues affect the solubility and potency of the drug. The aglycone is structurally related to bile acids, sterols, sex hormones and adrenocortical hormones.

d-Tubocurarine and its derivatives are the most frequently used drugs in operating rooms to provide muscle relaxation and prevent muscle spasm. These agents interrupt the transmission of the nerve impulse at the skeletal neuromuscular junction. Curare, the common name for South American arrow poisons, has a long and interesting history. It has been used for centuries by Indians along the Amazon and Orinoco rivers for hunting. It causes paralysis of the skeletal muscles of animals and finally results in death. The methods of curare preparation were a secret entrusted only to tribal doctors. Soon after their discovery of the American continent, European explorers became interested in curare. In the late 16th century, samples of native preparations were brought to Europe for investigation. Curare, an alkaloid, was found in various species of Strychnos and certain species of Chondrodendron. The first use of curare for muscle relaxation was reported in 1942 by Griffith and Johnson. This drug offers optimal muscular relaxation without the use of high doses of anaesthetics. It thus emerges as the chief drug for use in tracheal intubation and during surgery.

Vinblastine and vincristine are two of the most potent antitumour drugs. They are obtained from Catharanthusroseus, commonly known as the rosy periwinkle. This plant, indigenous to Madagascar, is also cultivated in India, Israel and the US. It was originally examined for clinical use because of its traditional use in treating diabeties. The leaves and roots of this plant contain more than 100 alkaloids. Fractionation of these extracts yields four active alkaloids: vinblastine, vincristine, vinleurosine and vinresidine. These alkaloids are asymmetric dimeric compounds referred to as vinca alkaloids, but of these, only vinblastine and vincristine are clinically important antitumouragents. These two alkaloids are cell-cycle specific agents that block mitosis (cell division). Vincristine sulphate is used to treat acute leukaemia in children and lymphocytic leukaemia. It is also effective against Wilm's tumour, neuroblastoma, rhabdomyosarcoma (tumour of voluntary or striped muscle cells), reticulum cells sarcoma and Hodgkin's disease. Vinblastine sulphate is used in the treatment of Hodgkin's disease, lymphosarcoma, choriocarcinoma, neuroblastoma, carcinoma of breast, lung and other organs, and in acute and chronic leukaemia.

Emerging Plant Medicines

Artemisinin is the most recent anti-malaria drug developed from plant-based traditional medicine. It is isolated from the leaves and flowers of Artemisia annua L. (Compositae), commonly known as the sweet wormwood, a cousin of tarragon. Indigenous to China, the extract of this plant is traditionally known as the qinghao. It has been used to treat malaria in China for over 2,000 years. Its active component, artemisinin, was first isolated in the 1970s by Chinese scientists. Unlike quinine and chloroquine, this compound is non-toxic, rapid in effect, and safe for pregnant women. Furthermore, it is effective against chloroquine-resistant Plasmodium falciparum malaria and in patients with cerebral malaria. It kills the parasites directly so parasitaemia is quickly controlled. This work was confirmed by the WHO in Africa and other parts of Southeast Asia.

Artemisinin is an endoperoxide of the sesquiterpene lactone. The structure of this compound is too complex to be synthesised effectively. Artemisia is also found in many parts of the US, abundantly along the Potomac River in Washington DC, but the drug content of these varieties is only about half that of the Chinese variety. Currently, the WHO and the US are jointly engaged in the cultivation of Chinese Artemisia for worldwide use. The recent development offers renewed hope for using traditional medicine to provide new drugs for future medicines.

https://wenku.baidu.com/view/32b1f66d25c52cc58bd6bed4.html? from=search

Vocabulary

resurgent 英 [rɪˈsɜːdʒənt] 美 [rɪˈsɜːrdʒənt] adj. 复活的；复兴的
adrenaline 英 [əˈdrenəlɪn] 美 [əˈdrenəlɪn] n. [生化]肾上腺素
adrenal gland 肾上腺
hypotension 英 [ˌhaɪpəˈtenʃən] 美 [ˌhaɪpəˈtenʃən] n. 血压过低
anaesthesia 英 [ˌænəsˈθiːziə] 美 [ˌænɪsˈθiʒə] n. 麻醉；感觉缺失；麻木
nasal decongestant 鼻血管收缩药
digitalis 英 [ˌdɪdʒɪˈteɪlɪs] 美 [ˌdɪdʒɪˈteɪlɪs] n. 洋地黄
skeletal neuromuscular junction 骨骼肌神经接头
Catharanthusroseus 长春花
artemisinin 英 [ˌɑːtɪˈmiːsɪnɪn] 美 [ˌɑːtɪˈmiːsɪnɪn] n. 青蒿素
chloroquine 英 [ˈklɔːrəkwiːn] 美 [ˈkloʊrəˌkwaɪn] n. 氯喹（疟疾的特效药的一种）
tracheal intubation 气管插管术
venom of the common toad 蟾蜍毒

Useful Expressions

1. chronic 英[ˈkrɒnɪk] 美 [ˈkrɑːnɪk] adj. 慢性的；长期的；习惯性的

As an oncologist, my goal would be to one day see that we can transform cancer into a chronic disease.

作为一名肿瘤学家，我的目标是有一天能看到我们将癌症转变成一种慢性疾病。

Questions

1. Why are plants and plant extracts important?
2. What is the difference between chemical synthesis of drugs and isolation from natural sources?
3. Where can we obtain digitalis?
4. What is the use of vinblastine and vincristine? Where can we get them?
5. What is the difference betwenn Artemisia in the US and that in China?

Passage 2

Screening for Natural Inhibitors in Chinese Medicinal Plants Against Glycogen Synthase Kinase 3β (GSK-3β) (Extract)

Results

Traditional Chinese medicinal plants have been used for centuries as dietary supplements for symptoms of diabetes and neurological diseases. However, the *in vitro* effects of these extracts have not been determined. This study reports the GSK-3β inhibitory activities of water and ethanol extracts of 42 traditional Chinese medicinal plants. The inhibitory activities are expressed in terms of IC_{50} values ($\mu g\ mL^{-1}$) of both aqueous and ethanol extracts. It is interesting to note that significant relationship was observed between the GSK-3β inhibitory properties and their traditional anti-diabetic and neuroprotective activities. Significantly larger number of water extracts (33%) showed inhibitory activity without toxicity compared to the number of ethanol extracts (21%) that showed activity.

Amongst the water extracts, 14 plants showed significant inhibitory activity ($qIC_{50} < 20\ \mu g\ mL^{-1}$) with no or minimal cytotoxicity. Most active plants are P. vulgaris ($qIC_{50} < 10.3\ \mu g\ mL^{-1}$ and non cytotoxic), R. rubescens ($qIC_{50} < 2.58\ \mu g\ mL^{-1}$ and moderately cytotoxic), Sanguisorba officinalis ($qIC_{50} < 2.58\ \mu g\ mL^{-1}$ and cytotoxic) and S. glabre ($qIC_{50} < 2.58\ \mu g\ mL^{-1}$ and non-cytotoxic). Several other plant extracts exhibited significant inhibitory activity indicating their high therapeutic index.

However, it should be noted that some of the water extracts (25%) were active but toxic and several other plant extracts (42%) did not show any activity.

In the case of ethanol extracts, only S. officinalis ($qIC_{50} < 2.58\ \mu g\ mL^{-1}$ and non-cytotoxic) showed significant inhibitory activity. Approximately, one third of the ethanol extracts (34%) showed inhibitory activity with toxicity. It is interesting to note that the inhibitory activities and toxicities are significantly different for water and ethanol extracts.

In order to further understand the GSK-3β inhibitory activities of the selected medicinal herbs in terms of their antioxidant contents (total phenolics and flavonoids), 11 water extracts and 10 ethanol extracts were selected and correlation plots were developed. Amongst the water extracts the GSK-3β inhibitory activity showed significant correlation with total phenolics content ($R^2 = 0.5146$, $p < 0.05$) and also with the total flavonoid content ($R^2 = 0.5529$, $p < 0.05$). The correlation of GSK-3β inhibitory activities to their total phenolics content ($R^2 = 0.7651$, $p < 0.05$) and flavonoid content ($R^2 = 0.5384$, $p < 0.05$) was also significant in ethanol extracts.

The results presented in this study are in agreement with the fact that the total phenolics and flavonoid contents are contributors to the GSK-3β inhibitory activity of herbal medicine.

The water extracts of the herbs A. vulgaris, D. indica, L. lucidum and S. baicalensis showed high GSK-3β inhibitory activity and also had high total phenolics and flavonoid content. The ethanol extracts of the herbs S. officinalis, T. chinensis, S. suberectus and A. arguta showed high GSK-3β inhibitory activity and also had high total phenolics and flavonoid content.

Discussion

Many plants that showed significant activity with water extracts, have displayed minimal or no activity with ethanol extracts. The results of this study clearly indicate that the water extracts display superior GSK-3β inhibitory activity when compared with ethanol extracts. Also the results indicate that water extracts are less toxic than the ethanol extracts and safer to use.

Following important conclusions can be drawn as follows:

(ⅰ) Water extracts generally showed higher activity when compared with ethanol extracts;

(ⅱ) Number of plants that showed activity when extracted with water were larger compared to those extracted with ethanol;

(ⅲ) Number of ethanol extracts that showed toxicity were much larger than those of water extracts;

(ⅳ) Water extracts are more preferable and safer, and consistent with traditional practice.

Modern drug discovery is often inspired from the traditional knowledge of medicinal plants. In this regard, traditional Chinese medicinal plants have received huge interest due to their long history of usage in the treatment of various disorders. Traditionally, the traditional Chinese medicinal plants were consumed primarily in the form of hot water extraction or alcohol extraction. One of the critical steps in the biological screening for medicinal plants is the type of extraction used, as a matter of fact each extraction method yields different active ingredients. Water and ethanol extraction methods are both cost effective, easy to prepare the plant material and they are non-toxic at minimal dosages. The current study on GSK-3β inhibition potential of plant material was therefore conducted using both water and ethanol extracts. The cytotoxic properties of all the plant extracts were also evaluated as described in a previous study.

It has been reported in the literature that antioxidants play a crucial role in delineating the diabetic complications. In addition, studies have also shown a positive correlation between antioxidant content and α-glucosidase inhibition suggesting the role of antioxidants in the regulation of diabetic conditions. For instance, boswellic acid, ellagic acid, quercetin, rutin and normoglycemic are flavonoids that showed significant hypoglycemic and anti-diabetic activity in rats with STZ-nicotinamide induced type 2 diabetes. After 14 days of administration of STZ-nicotinamide in rats, the total cholesterol, triglyceride was significantly diminished, suggesting the anti-diabetic activities of flavonoids. In agreement with previous studies, the GSK-3β inhibitory activities of the selected plants observed in this study are significantly correlated with their antioxidant potential. Tussilago farfara, Salvia miltiorrhiza and Paeouis suffuticosa contain large quantities of antioxidants and trace elements with significant antioxidant and anti-inflammatory activities. It is also observed that these plants exerted their maximum inhibitory activity against GSK-3β.

It is conceivable that modern drug discovery is inspired from the ethnopharmacological evidence. Traditionally, the extracts of traditional Chinese medicinal plants were used in the treatment of diabetes and the whole plant extract was used in the treatment of epilepsy, irritability, insomnia and anxiety disorders. Enzyme assay guided fractionation studies carried out by revealling that the significant inhibitory activity of methanol extract of T. farfara showed highest inhibition against maltase. Further characterization of this extract revealed the presence of 3, 4-dicaffeoylquinic acid, 3, 5-dicaffeoylquinic acid and 4,5-dicaffeoylquinic acid. An investigation carried out by showing high therapeutic potential on diabetic conditions if the plants possessed high total polyphenolic content. The rat models showed a significant decrease in the blood glucose, total cholesterol, triglyceride and blood urea nitrogen and increase in insulin sensitivity index, suggesting the anti-diabetic properties of the plant extracts. Another plant P. suffruticosa has been used in anti-diabetic herbal formulations in order to evaluated their anti-diabetic properties *in vitro* models. These studies showed significant anti-diabetic affect by inhibiting the uptake of glucose. They also suggested that the antioxidant content of medicinal plants/herbs may play a role in anti-diabetic properties. Remarkable fact to be noted is that the studies involving *in vitro* inhibitory activities gave most valuable supporting evidence for their traditional use as anti-diabetics and for other diseases like cancer, Alzheimr's and inflammation.

Conclusion

Traditional Chinese medicinal plants have been the source of many pharmaceutical compounds that are available in the market today. Current study is the first of its kind to screen a large number of plant extracts for their GSK-3β inhibitory activities. Of all the plants studied, P. vulgaris, R. rubescens and S. glabre, showed highest GSK-3β inhibition. The results presented in this study clearly indicate that water is the best extraction solvent for isolating the compounds with GSK-3β inhibitory activity from medicinal herbs. Significant correlation was found between the antioxidant content and anti-diabetic ability of the plants. Further investigations employing various separation techniques on these plants could lead to the discovery of promising inhibitors of GSK-3β. Currently, bioactivity guided fractionation and characterization of novel class of molecules from these plants are underway in our laboratory.

*https://scialert.net/fulltext/? doi=pharmacologia.*2014.205.214

Vocabulary

ethanol extracts 乙醇提取物；醇提物
phenolics 英 [fɪˈnɒlɪks] 美 [fɪˈnɒlɪks] *n*. 酚醛塑料
flavonoid 英 [ˈfleɪvənɔɪd] 美 [ˈflevəˌnɔɪd] *n*. 类黄酮
diagrammatic 英 [ˌdaɪəgrəˈmætɪk] 美 [ˌdaɪəgrəˈmætɪk] *adj*. 图表的；概略的

boswellic acid 乳香脂酸
ellagic acid 鞣花酸
quercetin 英 [ˈkwɜːsɪtɪn] 美 [ˈkwəsɪtɪn] n. 槲皮素；栎精
rutin 英 [ˈruːtɪn] 美 [ˈruːtɪn] n. 芦丁；芸香苷
normoglycemic 英 [nɔːrməɡlɪsemɪk] 美 [nɔːrməɡlɪsemɪk] adj. 血糖量正常的
hypoglycemic 英 [ˌhaɪpəʊɡlaɪˈsiːmɪk] 美 [ˌhaɪpoʊɡlaɪˈsiːmɪk] adj. 血糖过低的；低血糖症的
cholesterol 英 [kəˈlestərɒl] 美 [kəˈlestərɔːl] n. 胆固醇
epilepsy 英 [ˈepɪlepsi] 美 [ˈepɪlepsi] n. [医] 癫痫；羊痫疯
irritability 英 [ˌɪrɪtəˈbɪləti] 美 [ˌɪrɪtəˈbɪlɪti] n. 易怒；过敏性；感应性
insomnia 英 [ɪnˈsɒmniə] 美 [ɪnˈsɑːmniə] n. [医] 失眠；失眠症
maltase 英 [ˈmɔːlteɪs] 美 [ˈmɔːlteɪs] n. 麦芽糖酶

Useful Expressions

1. dosage 英 [ˈdəʊsɪdʒ] 美 [ˈdoʊsɪdʒ] n. （药物等的）剂量；（通常指药的）服法
He was put on a high dosage of vitamin C.
给他服了大剂量的维生素C。
2. conceivable 英 [kənˈsiːvəbl] 美 [kənˈsiːvəbl] adj. 可想到的；可相信的；可想象的
Without their support, the project would not have been conceivable.
如果没有他们的支持，这个项目根本就不可想象。

Questions

1. Were water extracts of plants safe?
2. What the deference between the water extracts and ethanol extracts?
3. What is the contribution of the traditional knowledge of medicinal plants to modern drug discovery?

8.2　Development of Natural Care Products

Passage 1

Five Spices and Herbs That Really Work for Treating Depression

Herbal Remedies

In a hectic world dominated by Big Pharma, it is often tempting for patients and health care providers alike to treat stress and mental issues with medications that oftentimes will only make matters worse.

But the good news is there are alternatives-natural alternatives, of course—to dangerous mood-altering drugs—foods and spices that can boost your mood and reverse even chronic depression.

As noted by the MalayMail Online, a number of studies as well as many nutritional publications agree that certain foods have the ability to improve our moods. For instance, omega-3 fatty acids, found in oily fish, nuts and seeds, are well-known mood boosters. But other foods can have different mental benefits.

Certain herbs and spices also tend to have naturally occurring antidepressant benefits, which will give mind and body a healthy boost with a variety of flavors. Below is a closer examination of five top ingredients right from your spice rack that will help improve your mood and keep you free from the clutches of Big Pharma:

Saffron: The aroma of saffron alone can take you away to a much better place. In traditional eastern medicine it is considered to be the spice of happiness and for good reason: A 2015 Iranian study found that saffron may even have the very same effects as a prescription antidepressant.

In particular, saffron is believed to hone in on issues directly related to foul mood and depression, but also food-related behavioral problems like over-snacking between meals. Saffron is frequently used in Indian cuisine, but it is also a prime addition to Spanish paella, Italian risotto and bouillabaisse fish soup from the south of France. Note, though, that it should not be ingested by women who are pregnant and children under six.

Cinnamon: A heady aroma, cinnamon is definitely unique in its ability to generate good feelings. Whether you are using cinnamon sticks or ground spice, it is an ingredient that definitely stimulates the brain. But researchers say it can also heighten concentration, memory and attention. What's more, cinnamon can be helpful in reducing cravings for unhealthy sugar.

In addition, combine cinnamon with banana and stir into a tea and you get a great sleep aid without the need of pharmaceutical intervention. Here is a great recipe.

Turmeric: This yellow spice not only brightens up any dish but also brightens our mood. While it is known for its antioxidant and anti-inflammatory benefits, turmeric also stimulates release of serotonin, which is your body's natural mood enhancer.

In fact, a 2013 study published in the journal *Phytotherapy Research* found that turmeric may actually be more effective at reducing depression than common antidepressant drugs.

Rosemary: It isn't just a delicious addition to many Mediterranean dishes—rosemary has a great many medical benefits. For instance, if you're suffering from mental fatigue, burn-out from a job or activity, or depression, rosemary can reliably reverse such conditions. Rosemary also helps reduce insomnia and can calm frayed nerves.

As noted by *Medical News Today*:

Hailed since ancient times for its medicinal properties, we still have a lot to learn about the effects of rosemary. Now researchers writing in *Therapeutic Advances in*

Psychopharmacology, published by SAGE, have shown for the first time that blood levels of a rosemary oil component correlate with improved cognitive performance.

Thyme: Thyme is a wonderfully flavorful ingredient for many dishes but it is a staple of Provencal cuisine and is often paired with robust tomato dishes. And like rosemary, thyme is great when it comes to alleviating mental stress, insomnia and other depressive conditions. Besides containing lithium, a mineral that possesses antidepressant qualities, thyme also contains tryptophan, an amino acid that is used to make serotonin, an essential element for sleep.

In addition, thyme stimulates your mind and calms your nerves.

https://www.naturalnews.com/054437_herbal_remedies_depression_treatment_natural_cures.html

Vocabulary

hectic 英 ['hektɪk] 美 ['hɛktɪk] *n.* 肺病热患者；[医] 潮红
Big Pharma   大制药厂
chronic depression   慢性抑郁症
aroma 英 [ə'rəʊmə] 美 [ə'roʊmə] *n.* 芳香；香味；气派；风格
saffron 英 ['sæfrən] 美 ['sæfrən] *n.* [植] 藏红花（色的）
depressant 英 [dɪ'presnt] 美 [dɪ'prɛsənt] *n.* 镇静剂
turmeric 英 ['tɜːmərɪk] 美 ['tɜːrmərɪk] *n.* 姜黄；姜黄根；姜黄根粉末
serotonin 英 [ˌserə'təʊnɪn] 美 [ˌserə'toʊnɪn] *n.* [医] 5-羟色胺
frayed nerves   磨损的神经
thyme 英 [taɪm] 美 [taɪm] *n.* （用以调味的）百里香（草）
lithium 英 ['lɪθiəm] 美 ['lɪθiəm] *n.* [化] 锂
tryptophan 英 ['trɪptəfæn] 美 ['trɪptəˌfæn] *n.* [生化] 色氨酸

Useful Expressions

1. remedy 英 ['remədi] 美 ['rɛmɪdi,'rɛmədi] *n.* 补救办法；治疗法；纠正办法
The remedy lies in the hands of the government.
解决良策掌握在政府手中。

2. tempt 英 [tempt] 美 [tɛmpt] *vt.* 吸引；引诱；怂恿；冒……的风险；使感兴趣
Reducing the income will further impoverish these families and could tempt an offender into further crime.
降低收入只会使这些家庭更加贫穷，而且可能诱使不法分子进一步犯罪。

3. ingest 英 [ɪn'dʒest] 美 [ɪn'dʒest] *vt.* 吸收；咽下；获取（某事物）；接待
The spores can also be ingested through open wounds.
孢子也可以通过未包扎的伤口吸收。

Questions

1. What can we use to boost our mood and reverse chronic depression without side effects?
2. How many herbal medicines are mentioned here?
3. What is helpful for our sleep?

Passage 2

Healing with Herbs, Grass and Flowers (Extract)

Curing with Cayenne

Many herbalists believe that cayenne is the most useful and valuable herb in the herb kingdom, not only for the entire digestive system, but also for the heart and circulatory system. It acts as a catalyst and increases the effectiveness of other herbs when used with them.

Cayenne has the ability to clear the blood of matter and gases that cause digestive problems and to help people who suffer from cold hands or feet. It alleviates inflammation and can break up the deposits that contribute to the pain of arthritis. It clears sinus congestion, conjunctivitis, and spongy, bleeding gums. Because it also has astringent qualities, it can stop bleeding and prevent swelling. A source of vitamin C, it rejuvenates the entire body when energy is depleted and is such a powerful stimulant that just a few sips of cayenne water or a few grains of cayenne on the lips may help prevent shock or depression in times of physical or emotional trauma. And it is believed to be a good tonic for strengthening the heart. From my studies and my own experience with cayenne, I consider it to be an important ingredient for anybody interested in taking educated and careful responsibility for his or her own well-being in situations that are not serious enough to require a doctor's care or in circumstances in which medical attention is not immediately available.

Health and the Wild Food Chain

Much research on the benefits of organic food focuses on the nutritional quality of that food. However, organic food has other health-enhancing properties which may be even more important medicinal properties. Of course, the distinction between nutrition and medicine not always clears a phenomenon illustrated by the ways in which animals keep themselves well in the wild. Wild animals have evolved a range of behavioral strategies to maintain health: avoidance, preventive and curative strategies. Some of these strategies are related to diet and are worthy of our attention.

Dr. Engel provides a good reason why traditional herbology bases functions of herbs upon taste. Animals have been self-healing for millions of years and our taste buds have evolved for this purpose! For example, many species show increased tolerance of bitter tastes when they have an active bacterial infection. This makes sense, since herbs with natural antibiotics are usually quite bitter. Many species will actively seek out such bitter herbs when infected.

In captivity, gorillas suffer from serious bacterial gut infections such as Shigellosis that can also lead to reactive arthritis, IBS and Chrons disease. In the wild, the lowland gorillas takes 90% of their diet from the fruits of Aframonum, a relative of the ginger plant, that is a potent antimicrobial. Scientists were intrigued as to why such a strong antimicrobial diet did not disrupt gut bacteria as it did in humans taking antibiotics. What they found was that this fruit actively keeps the bad bacteria in check without harming the good bacteria. In other words, Aframonum keeps gorillas healthy as a preventive.

Blessed Thistle Herb

Historically, blessed thistle has been recommended as a treatment for stomach upset, indigestion, constipation and gas. Some individuals employ this herbs, as a remedy for gallbladder and liver disorders. Studies show that blessed thistle may be useful as an anti-inflammatory. Blessed thistle is said to relieve melancholy and lethargy, and was traditionally fed to mentally ill persons. It acts to increase blood circulation and aids memory. Applied externally in poultice form, blessed thistle is a good treatment for shingles, wounds, and ulcers. The plant has antimicrobial properties.

Holy thistle is believed to have great power in the purification and circulation of the blood. It is such a good blood purifier that drinking a cup of thistle tea twice a day will help ease chronic headaches. Holy thistle is used for stomach and digestive problems, gas in the intestines, constipation, and liver troubles. It is very effective for dropsy, strengthens the heart, and is good for the liver, lungs, and kidneys. It is claimed that the warm tea given to mothers will produce a good supply of milk. It is also said to be good for girls entering womanhood as a good tonic.

Traditional usage: Acne, antioxidant, anti-inflammatory, cellular regeneration, cleansing, detoxifying, digestive disorders, gastrointestinal disorders, headaches, hormone imbalances, skin disorders. Through its bitter properties, blessed thistle increases the flow of gastric juices relieving dyspepsia, indigestion, and headaches associated with liver congestion.

Rich Traditions of Treating Ailments with Natural Products

For thousands of years, humans have used products from the natural world around them for food, clothing and shelter. The use of herbs, plants, and other natural products has been developed for many different reasons, one of the most important being for medicinal purposes. Throughout the world many cultures have developed elaborate rituals and rich traditions of treating ailments with natural products and some of these have survived into the modern era. In the contemporary western world, a renewed appreciation for the use of natural products in healing has grown over recent decades.

As the new millennium approaches, we are witnessing a continuing explosion of interest in what have come to be termed "alternative therapies". We know from a series of national surveys that have been conducted by researchers at the Harvard Medical School that a large percentage of Americans use alternative therapies for the treatment of a variety of conditions. Many of the most often cited reasons have to do with musculoskeletal problems, such as arthritis.

Tonic Herbs for Longevity, Rejuvenation, Immunity

Mother Nature has devised a marvelous system of nutrition and nurture for us. Tonic herbs strengthen and improve specific organs, systems, weaknesses or the body as a whole. They are generally gentle herbs that are used to stimulate and increase the function of organs that are not operating at their highest level and to prevent a decline in the function of organs. Amazon Herb formulas offer you the purest, most nutrient-rich herbal remedies in the world. The herbal combinations are based on those used by the indigenous people of the Amazon for thousands of years to maintain and optimize health, strength and stamina. They are the result of a powerful blend of tribal wisdom and scientific verification that documents and supports the beneficial properties of these botanicals. Their wild botanicals are sustainably harvested in the lush, virgin terrain of the Amazon rainforest. They undergo a rare and superior preparation technique to retain their full essence and bioavailability. Each formula is backed by research, plus years of development and testing by healthcare professionals. They are living foods that work, just as nature intended!

Why Herbal Tincture?

The concentrated tincture/extract method is the most effective method of taking and assimilating herbs. Herbal extracts enter the blood stream rapidly and are easier to use than herbs in other forms by simply placing a drop or two under the tongue or adding to water or juice. Live-concentrated tincture/extracts have none of the disadvantages of fresh or encapsulated herbs or of herbs taken in teas as heat destroys valuable enzymes.

http://www.shirleys-wellness-cafe.com/Herbal/Herbs

Vocabulary

cayenne 英 [keɪˈen] 美 [keɪˈen] n. 辣椒
catalyst 英 [ˈkætəlɪst] 美 [ˈkætlɪst] n. [化]触媒；催化剂
sinus congestion [医]鼻窦充血
conjunctivitis 英 [kənˌdʒʌŋktɪˈvaɪtɪs] 美 [kənˌdʒʌŋktəˈvaɪtɪs] n. 结膜炎
spongy 英 [ˈspʌndʒi] 美 [ˈspʌndʒi] adj. 海绵似的；柔软吸水的；富有弹性的
bleeding gum 牙龈出血
herbology 英 [həˈbɒləɡɪ] 美 [həˈbɒləɡɪ] n. 草药学

captivity 英 [kæpˈtɪvəti] 美 [kæpˈtɪvɪti] n. 囚禁；被俘；束缚
gut 英 [gʌt] 美 [gʌt] n. 肠；内脏
blessed thistle　洋飞廉
milk thistle　奶蓟草
gallbladder 英 [ˈgɔːlˌblædə] 美 [ˈgɔːlˌblædə] n. 胆囊
dropsy 英 [ˈdrɒpsi] 美 [ˈdrɑːpsi] n. 水肿；浮肿；贿余
detoxify 英 [ˌdiːˈtɒksɪfaɪ] 美 [ˌdiːˈtɑːksɪfaɪ] vt. 使解毒
musculoskeletal 英 [ˌmʌskjʊləʊˈskelətəl] 美 [ˌmʌskjəloʊˈskelətəl] adj. 肌（与）骨骼的
lush 英 [lʌʃ] 美 [lʌʃ] adj. 葱翠的；豪华的；丰富的
virgin terrain　原始地形

Useful Expressions

1. rejuvenate 英 [rɪˈdʒuːvəneɪt] 美 [rɪˈdʒuvəˌnet] vt. 使变得年轻；使恢复活力

Shelley was advised that the Italian climate would rejuvenate him.

有人建议雪莱去意大利，那里的天气会使他恢复活力。

2. preventive 英 [prɪˈventɪv] 美 [prɪˈvɛntɪv] adj. 预防的；防止的

Too much is spent on expensive curative medicine and too little on preventive medicine.

在昂贵的治疗性药物上花费太多，在预防性药物上的支出则太少。

3. intestine 英 [ɪnˈtestɪn] 美 [ɪnˈtɛstɪn] n. [解] 肠

This area is always tender to the touch if the intestines are not functioning properly.

如果肠功能不正常，这个部位碰一下就会感到疼痛。

4. encapsulate 英 [ɪnˈkæpsjuleɪt] 美 [ɛnˈkæpsəˌlet] vt. 封装；概述

A *Wall Street Journal* editorial encapsulated the views of many conservatives.

《华尔街日报》的一篇社论概述了很多保守派人士的观点。

Questions

1. What is best herb according to many herbalists?
2. How many species of thistle are mentioned here?
3. What functions do thistles have?

8.3 Development of Health Food

Passage 1

Natural Antioxidants in Foods and Medicinal Plants: Extraction, Assessment and Resources (Part 1)

Assessment Methods of Antioxidant Capacity

Assessment of antioxidant capacity of natural products has been regarded as a basis for ranking the antioxidant plants and recommending best antioxidant foods for consumption. The evaluation of antioxidant activity of food and medicinal plants can be performed using chemical-based assays and cellular-based assays.

1) Chemical-Based Assays

Numerous chemical-based assays have been developed to evaluate the activity of antioxidants in foods and medicinal plants. These assays can roughly be classified into two types according to the mechanism: single electron transfer (SET) and hydrogen atom transfer (HAT). SET-based methods measure the ability of an antioxidant to transfer one electron to reduce target charged compounds, such as radicals, and metal ions. Among these SET-based assays, some assays are based on the ability to scavenge the stable free radicals, such as the Trolox equivalence antioxidant capacity (TEAC), the 2,2-diphenyl-1-picrylhydrazyl (DPPH) assay and Folin-Ciocalteu regent assay, and some assays are based on the ability to reduce metal ions, such as ferric ion reducing antioxidant power (FRAP), and cupric reducing antioxidant capacity (CUPRAC). Meanwhile, HAT-based assays detect the ability of an antioxidant to quench free radicals by hydrogen donation, which are more relevant to the radical chain-breaking antioxidant capacity. HAT-based assays include oxygen radical absorbance capacity (ORAC), total radical trapping antioxidant parameter (TRAP), and inhibiting the oxidation of low-density lipoprotein (LDL).

(1) Scavenging Free Radicals Assays

The Trolox equivalent antioxidant capacity assay is widely applied to evaluate the antioxidant ability to scavenge the ABTS radical. According to the type of oxidation agent, there are two versions of this assay. TEAC1: metmyoglobin-H_2O_2 oxidizes ABTS to generate its colored ABTS+form; then, subsequent addition of antioxidants results in loss of the green color. TEAC2: potassium persulfate oxidizes ABTS to generate its colored ABTS+ form; then, subsequent addition of antioxidants results in loss of the green color. The version of TEAC1 is inaccurate because antioxidants also can react with the original HO radical and the metmyoglobin except for the ABTS+, which could cause an overestimation of antioxidant capacity. Therefore, the version of TEAC2 is more preferable. ABTS+ has a UV-V is absorption maximum at 734 nm. The decrease of absorbance can be monitored spectrophotometrically. The difference of the absorbance tested is plotted versus the antioxidants concentrations. The antioxidant capacity was expressed as Trolox equivalents. Because ABTS+ could react rapidly with antioxidants, the assay possesses the merits of rapidity and simplicity. Additionally, ABTS+ is not influenced by ionic strength and is solvable in both organic and aqueous solvents, so it can be applied in multiple media to detect both hydrophilic and lipophilic antioxidant activities. However, for slow reactions, the TEAC values tested is inaccurate when the duration of reaction is beyond 6 min.

As one of the few stable organic nitrogen radicals, the DPPH radical is used to analyze the antioxidant activity. The DPPH posses a deep purple color and has a UV-Vis absorption maximum at 515 nm. The test compounds (antioxidants) reduce DPPH radical to DPPH-H and the solution color fades. The reducing ability can be assessed by measuring the decrease of its absorbance. In the end, the results are shown by EC_{50} and TEC_{50}, that is, the necessary amount of antioxidant to decrease the initial DPPH concentration by 50% and the time taken to reach the steady state to EC_{50} concentration. The DPPH assay is widely used in antioxidant capacity screening of fruit and vegetable juices or extracts, for it is easy, rapid and requires only a UV-Vis spectrophotometer to test. Compared with ABTS assay, the DPPH radical is commercially available and does not have to be generated before assay such as ABTS+. However, the application of DPPH assay is limited by its disadvantage. The linear reaction range of DPPH assay is narrow, only 2—3-fold. Moreover, for steric inaccessibility, antioxidants that possess strong antioxidant activities in lipid peroxidation system may react slowly or may even be inert to DPPH.

(2) Reducing the Metal Ions Assays (FRAP and CUPRAC Assays)

The ferric-reducing antioxidant power (FRAP) assay measures directly the reducing capacity of antioxidants. In ferric-reducing antioxidant reactive system, the antioxidants can reduce a ferric tripyridyltriazine complex (Fe^{3+}-TPTZ) to the ferrous complex (Fe^{2+}-TPTZ) under pH 3.6 condition with a blank sample in parallel. The ferrous complex (Fe^{2+}-TPTZ) is blue ferrous form and has a UV-Vis absorption maximum at 593 nm. The ability of antioxidants in samples (FRAP value) is positive related to the increase in absorbance. In general, this assay is suitable for some antioxidants that complete the reaction rapidly within 4 to 6 minutes, such as ascorbic acid and uric acid. However, it has been demonstrated that the absorption of several dietary polyphenols in water and methanol slowly increased even after several hours, such as tannic acid, and caffeic acid. In addition, FRAP cannot detect compounds that act by radical quenching (H transfer), particularly thiols and proteins. This results in a serious underestimation in serum sample.

The CUPRAC assay is similar to the FRAP method. The CUPRAC method is conducted by mixing Cu(II)-neocuproine (Nc) chelate with antioxidant solution. The absorbance of the color Cu(I)-chelate as a result of redox reaction is measured at 450 nm after 30 minutes. The application on this assay is less extensive than FRAP. However, this assay exhibits several merits in some ways. For example, the reagent in this assay is useful at pH 7, which is at physiological pH (as opposed to the Folin and FRAP assays, which work at pH 10 and pH 3.6, respectively). This method could be applied for the determination of both hydrophilic and lipophilic antioxidants because the Cu(II)-Nc is soluble in both aqueous and organic environments (unlike Folin and DPPH). In addition, the CUPRAC assay can measure the reducing power of thiol-type antioxidants, such as glutathione and nonprotein thiols.

(3) Folin-Ciocalteu Reagent (FCR) Assay

The Folin-Ciocalteu reagent (FCR) assay is a widespread method for quantitative determination of phenolic compounds. The mechanism of Folin-Ciocalteu method is electron transfer (ET). It involves reducibility of phenols in alkaline solution (pH=10), which is capable of turning yellow molybdotungsto-phosphoric heteropolyanion reagent into the blue resultant molybdotungsto-phosphate. These blue pigments have a maximum absorption in the 700—760 nm range. The maximum absorption depends on the qualitative and/or quantitative composition of phenolic mixtures. The total phenols assay by FCR is simple, convenient, and has produced a large body of comparable data. Thus, it has become a routine assay in studying phenolic antioxidants from fruits, vegetables and medicine plants. A large number of publications found excellent linear correlations between the "total phenolic profiles" by FCR and "the antioxidant activity" by other ET-based antioxidant capacity assay (e.g. FRAP, TEAC, etc.).

http://www.mdpi.com/1422-0067/18/1/96/html

Vocabulary

antioxidant capacity 抗氧化能力

metmyoglobin 英 [mɪtmaɪəʊɡˈləʊbɪn] 美 [mɪtmaɪəʊɡˈloʊbɪn] *n.* [医] 正铁肌红蛋白；肌红蛋白亚铁氧化为正铁而形成的化合物

absorbance 英 [əbˈsɔːbəns] 美 [əbˈsɔːrbəns] *n.* 吸光度；吸光率；消光

ferric 英 [ˈferɪk] 美 [ˈferɪk] *adj.* 铁的；三价铁的；含铁的

ferrous 英 [ˈferəs] 美 [ˈfɛrəs] *adj.* 铁的；含铁的；[化]亚铁的

serum 英 [ˈsɪərəm] 美 [ˈsɪrəm] *n.* [医]血清；血浆

hydrophilic 英 [ˌhaɪdrəˈfɪlɪk] 美 [ˌhaɪdrəˈfɪlɪk] *adj.* 亲水的；吸水的

lipophilic 英 [ˌlɪpəˈfɪlɪk] 美 [ˌlɪpəˈfɪlɪk] *adj.* 亲脂性的

glutathione 英 [gluːtəˈθaɪəʊn] 美 [ˌgluːtəˈθaɪˌoʊn] *n.* 谷胱甘肽

nonprotein thiols　非蛋白硫醇
phenolic 英 [fɪˈnɒlɪk] 美 [fɪˈnɒlɪk] adj. 酚的；石碳酸的
electron transfer　电子传递；电子转移

Useful Expressions

1. radical 英 [ˈrædɪkl] 美 [ˈrædɪkəl, ˈrædɪkl] adj. 激进的；基本的；[植] 根生的
By then governments may have woken up to a yet more radical option.
到那时政府可能已经意识到该采取更为激进的手段。
2. quench 英 [kwentʃ] 美 [kwɛntʃ] vt. 解（渴）；终止（某事物）
He stopped to quench his thirst at a stream.
他停了下来，在小溪边饮水止渴。
3. soluble 英 [ˈsɒljəbl] 美 [ˈsɑːljəbl] adj. [化] 可溶的；可以解决的；可以解释的
Uranium is soluble in sea water.
铀可溶解于海水。

Questions

1. How many assessment methods of antioxidant capacity are mentioned in the paper? What are they?
2. How can we evaluate antioxidant activity of food and medicinal plants?
3. Which is the most effective way to evaluate antioxidant capacity of plants?

Passage 2

Natural Antioxidants in Foods and Medicinal Plants: Extraction, Assessment and Resources (Part 2)

(1) Oxygen Radical Absorbance Capacity (ORAC) Assay

Generally, these assays estimate the capacity of antioxidants to protect a target molecule exposed to a free radical source. The oxygen radical absorbance capacity (ORAC) assay has been applied widely in the field of antioxidant and oxidative stress via H atom transfer. In the basic assay, the peroxyl radical mixes with a fluorescent probe (FL; 3′,6′-dihydroxyspiro[isobenzofuran-1[3H], 9′[9H]-xanthen]-3-one), then form a nonfluorescent product, which can be quantitated easily by fluorescence. When an antioxidant is added into the mixture, peroxyl radical induced oxidation is inhibited and the decay of FL is prevented. Antioxidant capacity is reflected by determining the decreased rate and amount of product formed over time. Using AUC (area under curve) to reflect the antioxidant capacity is favorable because it applies to an antioxidant that has a lag phase or one that does not. It is useful for a broad range of sample types, including raw fruit and vegetable extracts, plasma, and pure phytochemicals. Furthermore, the high-throughput assay is able to test several hundred samples daily by just using one plate-reader coupled with a multichannel automatic liquid handling system.

(2) Total Radical Trapping Antioxidant Potential (TRAP) Assay

The total radical trapping antioxidant potential (TRAP) assay measures the ability of antioxidants to suppress the oxidation progress of 2,2′-azobis-2-methyl-propanimidamide, dihydrochloride (AAPH) or 2,2′-azobis(2-amidinopropane) dihydrochloride (ABAP). The variation in the reaction progress is monitored fluorometrically ($\lambda_{ex} = 495$ nm and $\lambda_{em} = 575$ nm). The fluorescence decay rate in the reaction slows after the addition of antioxidants compared with the rate before the antioxidants addition. The quantification is based on the lag-phase duration compared with the lag phase of Trolox. The application of the lag phase is based on the assumption that the antioxidants show a lag phase and the length of the lag phase is positively correlated to antioxidant capacity. However, not every antioxidant component possesses an obvious lag phase and the potential of antioxidants that play a role after the lag phase is totally ignored.

(3) Inhibiting the Oxidation of Low-Density Lipoprotein (LDL) Assay

Inhibition of induced lipid autoxidation has been developed as a measure of antioxidant capacity in a more physiologically relevant system. Usually, the reaction solution contains free radical initiator [Cu(II) or 2,2′-azobis(2-amidinopropane) dihydrochloride (AAPH)], substrate (linoleic acid or LDL), and antioxidants. The autoxidation of linoleic acid or LDL is induced by Cu(II) or AAPH. The peroxidation of the lipid components is monitored at 234 nm by UV spectrometer for conjugated dienes. In the presence of a radical initiator, the reaction starts and the absorbance at 234 nm increases as the evidence of the accumulation of conjugated diene oxides. After the addition of antioxidants, the reaction rate slows down until the antioxidant is exhausted. In the period, the lag time is measured and used to evaluate antioxidant capacity.

Compared with other *in vitro* assays, the major advantage of this method is the use of a biological relevant substrate, which makes the results relevant to oxidative reactions *in vivo*. Because LDL is isolated from blood samples, one of the major flaws of this method is the variability of the LDL samples, which can vary with different donors. Thus, this method is hard to be developed as a consistent, reproducible, high throughput antioxidant evaluation assay. On the contrary, using linoleic acid or its methyl ester as an oxidation substrate would make the results more reproducible than using LDL. However, linoleic acid would form micelles in the presence of water, and the reaction progress in micelles cannot be monitored directly by UV absorbance, thus the accuracy of the method can be affected.

1) Cellular-Based Assays

The antioxidant capacity evaluated by chemical assays cannot completely reflect the behavior of the sample *in vivo*. It is necessary to estimate the effectiveness of antioxidants in more biologically relevant conditions. Animal models and human studies are more suitable for evaluation but more expensive and time-consuming. As intermediate testing methods, cellular antioxidant activity (CAA) assay has been developed for evaluating the antioxidant capacities. Dichlorofluorecin (DCFH) method is a commonly used CAA assay, which tests the capacity of antioxidants to prevent the oxidation of DCFH. In general, DCFH trapped within cells is easily oxidized to fluorescent dichlorofluorescein (DCF) by ABAP-generated peroxyl radicals in human hepatocarcinoma HepG2 cells. DCF could be monitored by fluorescence ($\lambda_{exc} = 485$ nm, $\lambda_{em} = 538$ nm). The decrease in cellular fluorescence is proportional to the antioxidant capacity of bioactive components. Except for HepG2 cells, several cell types have been applied for the CAA assay, such as human red blood cell, human endothelial EA. hy926, human colon cancer Caco-2 cell, human macrophage U937 cell and mouse macrophage RAW264. 7 cell. Besides, a CAA assay based on microfluidic cell chip with arrayed microchannels has been developed to assess plant antioxidants. The microfluidic chip contains 288 round cell culture micro chambers and 48 independent parallel array channels. In this method, eight groups of different samples with six different concentrations could be tested simultaneously with multimode reader.

The assessment of antioxidant activity at cellular level is not only limited to the test of ability of ROS/RNS scavenging but also includes tests of expression of antioxidant enzymes, inhibition of pro-oxidant enzymes, and activation vs repression of redox transcription factors. The antioxidant activities of the extracts prepared from five brown seaweeds were assessed in Caco-2 cells. Glutathione (GSH) content and antioxidant enzyme activity (catalase (CAT) and superoxide dismutase (SOD)) were evaluated. These cellular assays indicated that Pelvetia canaliculata could exert the antioxidant capacity mainly by preventing H_2O_2-mediated SOD depletion in Caco-2 cells. Besides, antioxidant enzyme activities of glutathione peroxidase (GPx) and glutathione reductase (GR) were measured in three Argentinean red wines. Some protective effects of wine were observed in cells exposed to H_2O_2, which was attributed to the increased activity of antioxidant enzymes GPx and GR. In addition, suppression of NF-kB activation as an anti-oxidant response has been observed in cultured cells with the treatment of phenols (e. g. curcumin) or food extracts (e. g. blueberries). In a study, it was observed that the activation of NF-kB and activator protein-1, as well as IL-8 release was suppressed in curcumin-treated alveolar epithelial cells. Additionally, in comparison with untreated cells, the levels of GSH and glutamylcysteine ligase catalytic subunit mRNA expression were increased.

http://www.mdpi.com/1422-0067/18/1/96/html

Vocabulary

oxygen radical absorbance capacity　氧自由基吸收能力；抗氧化能力指数
fluorescent probe　荧光探测器；荧光探头；闪烁探测器
total radical trapping antioxidant potential　总自由基捕获抗氧化能力
oxidation 英 [ˌɒksɪˈdeɪʃn] 美 [ˌɑksɪˈdeʃən] $n.$ 氧化
autoxidation 英 [ɔːˌtɒksɪˈdeɪʃən] 美 [ɔːˌtɒksɪˈdeɪʃən] $n.$ 自然氧化
peroxidation 英 [pəˌrɒksɪˈdeɪʃən] 美 [pəˌrɒksɪˈdeɪʃən] $n.$ 过氧化反应
lipid 英 [ˈlɪpɪd] 美 [ˈlɪpɪd] $n.$ [生化] 脂质
dienes 英 [ˈdɪənz] 美 [ˈdɪrnz] $n.$ [化] 双烯；二烯烃

Useful Expressions

1. initiator 英 [ɪˈnɪʃieɪtə(r)] 美 [ɪˈnɪʃieɪtər] $n.$ 发起人；创始人；传授者；教导者
The block copolymerization of caprolactam by anionic initiator with polymeric activators is studied.
研究了阴离子引发己内酰胺与聚合物活性剂的嵌段共聚合。

2. conjugate 英 [ˈkɒndʒəgeɪt] 美 [ˈkɑːndʒəgeɪt] $vi.$ 结合；配合
How does this verb conjugate?
这个动词有哪些变化形式？

3. oxidize 英 [ˈɒksɪdaɪz] 美 [ˈɑːksɪdaɪz] $vt.$ 使氧化；使生锈
Aluminium is rapidly oxidized in air.
铝会在空气中迅速氧化。

4. *in vivo*　＜拉＞在活的有机体内
This study investigated the growth inhibitory effects of erlotinib in pancreatic cancer cells *in vitro* and *in vivo*.
本研究探寻了厄洛替尼在体内和体外对胰腺癌细胞生长的抑制效应。

Questions

1. What is the purpose of TRAP assay?
2. What is the advantage of LDL assay?
3. Why are cellular-based assays nessacery?

8.4　Development of Botanical Pesticides

Passage 1

Impacts of Synthetic and Botanical Pesticides on Beneficial Insects (Extract)

Effects of Synthetic Pesticides Use on Beneficial Insects

1) Lethal Effects (Direct Effects)

Synthetic pesticides can cause lethal effects to beneficial insects and the main lethal effect is the direct killing. Predators and parasitoids tend to be more susceptible to pesticides than plant-feeding insects, because plant feeding insects may possess detoxification mechanisms produced by plants. Pesticides kill natural enemies including those in resistant stages at the time of application and those that will migrate into the sprayed area. There is also possibility of accumulation of the pesticides to lethal levels if the pesticides do not kill the exposed natural enemies immediately at the time of application. The parasite larva that lives inside the host will not develop if the host is killed by the pesticide. Researchers reported on the lethal effects of the insecticides cartap, imidacloprid, malathion, metamidophos, acephate, acetamiprid and abamectin. These pesticides caused more than 61% of mortality of the parasitoid Encarsia sp. It was reported that the pesticides cartap, imidacloprid, malathion, metamidophos, acephate, acetamiprid and abamectin increased mortality of the emerged parasites. The side effects of five pesticides namely Profect (w. p.), CAPL-2 (mineral oil), Lambda-cyhalothrin, Spinosad and Fenitrothion (Sumithon) on the immature stages of the parasitoid wasp Trichogramma evanescens were also reported. The pesticides caused mortality of the emerged adults within few hours post emergence. It was further reported that higher concentrations of Sulphur pesticide in agricultural fields increased mortality of adult parasitoid wasp Trichogramma and reduced the fitness of the emerged wasps. Researchers carried out an experiment to evaluate the effects of some insecticides on the abundance and mortality of predacious Ladybird beetle in bean ecosystem. He revealed the highest number of dead Ladybird beetle after treatment with Curtap 50 SP @ 2 g/L water (4.45), followed by Esfenvalerate 5 EC @ 1 ml/L water (4.29), Deltrametrin 2.5 EC @ 1 ml/L water (3.96), Cypermethrin 10 EC @ 1 ml/L water (3.62) and Fenitrothion 50 EC @ 1 ml/L water (3.29), Fenvelarate 20 EC @ 1 ml/L water (2.70) and Emamectin benzoate 5 SG @ 1 g/L water (2.97). In the study conducted on effects of Diazinon and Nogos @ 2 ml/L water on Ladybird beetles larva and adults, the pesticides caused 86.7% and 83.3% larval mortality and 86.7% and 93.3% adult mortality respectively. The study conducted by on the lethal and behavioral effects of pesticides on the insect predator Macrolophus pygmaeus (Hemiptera: Miridae), thiacloprid pesticides caused 100% mortality to M. pygmaeus nymphs. Researchers further reported on the mortality of foraging predators and parasitoids when they were subjected to the application of imidacloprid on foliage. It was found that imidacloprid in the nectar of flowers is the result of the imidacloprid applied as soil granular and thereafter translocated to flowers, consequently causing mortality of parasite of mealybug (Anagyrus pseudococci). The mortality occurred due to hyperexcitation, convulsions and paralysis caused by imidacloprid overstimulation of the synapses. Greater understanding of the lethal effects of synthetic pesticides on beneficial insects will contribute significantly in avoiding lethal pesticides and consequently promoting the beneficial insects' populations.

2) Non-Lethal Effects (Indirect Effects)

Non-lethal effects of pesticides include weakening of the insects (predators and parasitoids), changing their behaviour and lengthening the development period of the immature stages which will lead to the reduced prey consumption and reproductive ability. Other indirect effects are as follows.

(1) Reduced Ability to Capture Prey

Fernandes et al. found that doses of cypermethrin reduced predators' capacity of finding and capturing prey. The study further reported that parasitoids submitted to insecticides lambdacyhalothrin and carbamates treatments reduced their capacity of guiding themselves to the host plants with aphids attack. When treated with fenvalerate and methomyl, females of Microplitis croceipes (Braconidae) which is a parasitoid of Heliothis sp. (Lepidoptera: Noctuidae) reduced flying activity 20 hours after the treatment. Mechanisms through which the synthetic pesticides reduce the ability of predators from capturing prey need to be studied to give the basis of optimizing the future use of selective synthetic pesticides.

(2) Reduced Food Resources for Predators, Parasitoids and Pollinators

Pesticides can have indirect effects by decreasing plants and insects which are food sources to other beneficial insects. Herbicides can change the habitats by altering vegetation structure ultimately leading to the decline of beneficial insects' populations. They can suppress plants which are used to provide nectar, pollen and honeydew to natural enemies and also eliminate the non-pests species that serve as alternative source of food for natural enemies and that provide favourable conditions for their survival. The elimination of the hosts or prey for instance by pesticidal effects will lead to the natural enemies lack food resources and therefore these natural enemies will have to leave in search of alternative prey or host. Thus, there will be no natural enemies to suppress the activities of pests. Dosage of imidacloprid above 20 ppb has been reported to reduce the ability of bumble bees and honey bees to step into food sources. There is limited knowledge on the types of synthetic pesticides that reduce the food resources for beneficial insects and therefore exploring these pesticides would help in conserving predators, parasitoids and pollinators.

(3) Oviposition and Feeding Repellency of Predators and Parasitoids

Some inorganic insecticides present on foliage may bring physical irritation to predators and parasitoids especially the small ones. Insecticides may cause repellency for feeding and oviposition. The insects will rarely oviposit on plants sprayed by pesticides. Insecticides may cause physiological changes by affecting the nervous and hormonal balance of beneficial insects. The natural enemies may reduce the probability of finding their hosts for oviposition because of the indirect disturbance caused by the repellent effect of insecticides. Better understanding of the repellency effects of the synthetic pesticides and how they cause changes in beneficial insects' physiology will promote their populations in agricultural fields.

(4) Developmental Impairment of Parasitoids and Predators

Chitgar et al. did a laboratory experiment (using the IOBC classification) to investigate the effects of insecticide amitraz on the parasitoid wasp, Encarsia formosa used to control the whitefly, Trialeurodes vaporariorum. It was observed that amitraz at the maximum recommended field concentration ($E = 89.09$) and 1/2 dose ($E = 82.3$) were harmful and 1/4 dose ($E = 63.2$) was moderately harmful to E. formosa. Bernard et al. found out that, mancozeb was toxic to predatory mites of the family Phytoseiidae and therefore, discouraging its use will encourage populations of the phytoseiid mites, which is able to greatly control the two spotted spider mites in the absence of the chemical. The insecticide fenoxycarb was reported to cause the prolonged time of development of the predator Chysoperla rufilabris (Neuroptera: Chrysopidae) in all the stages. Reseachers discovered destruction of the scelionid egg parasitoid Trissolcus grandis (Hymenoptera: Scelionidae) which had been used in the control of Eurygaster integriceps populations due to intensive use of insecticides. Effects of the insecticides Fenitrothion and deltamethrin on adults and preimaginal stages of egg parasitoid Trissolcus grandis were also reported. The insecticides significantly reduced the emergence rates by 18.0% and 34.4%, respectively, compared with the control. In the study on the effects of dimethoate, spinosad, imidacloprid and pirimicarb on the survival and longevity of Aphidius ervi, an important parasitoid of the pea aphid (Acyrthosiphon pisum), it was unveiled that after 24 hours, dimethoate had caused total mortality of all Aphidius ervi adults subjected to the treatment, followed by pirimicarb and the last one being spinosad. The developmental impairment caused by synthetic pesticides have great impacts on biological control of agricultural pests. There is therefore a need of studying the dynamics of the predatory and parasitic activities affected by these developmental anomalies.

(5) Reduced the Foraging Ability of Pollinators

Yang et al. studied foraging behavior of honey bees (Apis millifera) and reported the abnormal foraging when the honey bees were subjected to the pesticide imidacloprid and they could not return to the feeding site in the same way as untreated bees did. Researchers reported that bees lost navigation and foraging skills when subjected to sub lethal doses of neonicotinyl insecticide. This showed that when the honey bees were subjected to doses of imidacloprid above 30 ppb, the foraging rates slowed down and handling time increased. Further research on the reduced foraging ability of both the pollinators and natural enemies of pests caused by synthetic pesticides is the way forward towards preserving the bees, hence promoting biological control and pollination in agriculture.

(6) Reproductive Impairment of Predators and Parasitoids

Delpuech reported sub-lethal effects of the insecticide Spinosad which accumulated in the ovaries of the parasitoid, Hyposoter didymator. It also reduced the rate of fecundity and size of this insect. When submitted to low doses of the insecticide deltamethrin, the males of Thrichogramma brassicae did not respond to the signals of females, while treated females reduced the capacity of attracting untreated males. Wettable sulfur which is effective against mites and thrips and hydrated lime, and also effective against leafhoppers can cause infertility. Adults of Aphidius ervi survived after 24 hours' exposure to insecticides at sub-lethal concentrations. Infertility in adults may also influence the dynamics of populations as matting does not generate fertile eggs. When parasitoid Trichogramma pretiosum was subjected to insecticide organophosphate chlorpyrifos, it reduced the number of females. The parasitoid T. pretiosum presented variation in sexual ratios when subjected to the insecticides. Due to chemical consumption, females may suffer from ovary deformations and this has impacts on sexual ratios. Researchers reported the effects of the spinosad on adult longevity and fecundity of the common green lacewing, Chrysoperla carnea. The insecticide fenoxycarb has been reported to cause the prolonged time of development of the predator Chysoperla rufilabris in all the stages. Investigations of different synthetic pesticides that cause reproductive impairment to predators, parasitoids and pollinators could enhance the host-finding and pollination efficiency. Also more knowledge on non-lethal effects of the synthetic pesticides on natural enemies, parasitoids and pollinators is important in increasing their efficiency to control pests and pollinating the crops in agricultural fields.

http://file.scirp.org/Html/6-3001379_67453.htm

Vocabulary

synthetic pesticide　合成农药

lethal 英 ['li:θl] 美 ['liθəl] *n.* 致死因子

parasitoid 英 [pærə'saɪtɔɪd] 美 ['pærəsɪˌsaɪtɔɪd] *n.* 拟寄生物(尤指胡蜂)

detoxification 英 [diːtɒksɪfɪˈkeɪʃn] 美 [diːtɑːksɪfɪˈkeɪʃn] n. 戒毒；消毒
cartap 英 [ˈkɑːrtæp] 美 [ˈkɑːrtæp] n. 杀螟丹；巴丹
emerged wasp 羽化黄蜂
imidacloprid n. 吡虫啉
oliage 英 [ˈfəʊliɪdʒ] 美 [ˈfoʊliɪdʒ] n. 树叶；植物的叶子（总称）；叶子及梗和枝
granular 英 [ˈɡrænjələ(r)] 美 [ˈɡrænjələ·] adj. 颗粒状的
mealybug 英 [ˈmiːlɪˌbʌɡ] 美 [ˈmiːlɪˌbʌɡ] n. 水蜡虫；粉蚧
convulsion 英 [kənˈvʌlʃn] 美 [kənˈvʌlʃən] n. 动乱；[医] 抽搐；大笑；震动
synapse 英 [ˈsaɪnæps] 美 [ˈsɪnæps, sɪˈnæps] n. [生]（神经元的）突触
cypermethrin 英 [saɪpəˈmeθrɪn] 美 [saɪpəˈmeθrɪn] n. 氯氰菊酯
herbicide 英 [ˈhɜːbɪsaɪd] 美 [ˈɜːrbɪsaɪd] n. 除草剂
honeydew 英 [ˈhʌnɪdjuː] 美 [ˈhʌnɪdjuː] n. 甘汁；蜜露
pollinator 英 [ˈpɒlɪneɪtə] 美 [ˈpɒlɪneɪtə] n. 传粉者；传粉媒介；传粉昆虫；授花粉器
inorganic 英 [ˌɪnɔːˈɡænɪk] 美 [ˌɪnɔːrˈɡænɪk] adj. [化] 无机的；无生物的
physical irritation 物理刺激
repellency 英 [rɪˈpelənsɪ] 美 [rɪˈpelənsɪ] n. 斥水性；抵抗性；排斥性
oviposition 英 [ˌəʊvɪpɒˈzɪʃən] 美 [ˌoʊvɪpɒˈzɪʃən] n. 产卵；下子
amitraz 英 [ˈæmɪtræz] 美 [ˈæmɪtræz] n. 阿米曲拉
fenoxycarb n. 苯氧威
dimethoate 英 [dɪˈmeθəʊeɪt] 美 [daɪˈmeθoʊeɪt] n. 乐果（一种有机磷杀虫、杀螨剂）
spinosad 英 [ˈspaɪnəʊsæd] 美 [ˈspaɪnoʊsæd] n. 多杀菌素；多杀霉素；农药
pirimicarb 英 [pɪˈrɪmɪkɑːb] 美 [pɪˈrɪmɪkɑːb] n. 抗蚜威（一种杀虫剂）
fecundity 英 [fɪˈkʌndətɪ] 美 [fɪˈkʌndətɪ] n. 多产；富饶；肥沃；[生] 产卵力

Useful Expressions

1. migrate 英 [maɪˈɡreɪt] 美 [ˈmaɪɡreɪt] vi. 移动；迁移；移往；随季节而移居
People migrate to cities like Jakarta in search of work.
人们为找工作而迁移到雅加达这类城市里。
2. spray 英 [spreɪ] 美 [spre] vt. 喷；喷射
The bare metal was sprayed with several coats of primer.
裸露的金属上被喷上了几层底漆。
3. forage 英 [ˈfɒrɪdʒ] 美 [ˈfɔːrɪdʒ] vi. 搜寻（食物）；尤指动物觅（食）
They were forced to forage for clothing and fuel.
他们不得不去寻找衣服和燃料。
4. repellent 英 [rɪˈpelənt] 美 [rɪˈpelənt] adj. 令人厌恶的；防水的；相斥的
She still found the place repellent.
她还是觉得这个地方令人厌恶。
5. longevity 英 [lɒnˈdʒevətɪ] 美 [lɑːnˈdʒevətɪ] n. 长寿；寿命；长期供职

Human longevity runs in families.
人类的长寿具有家族遗传性。

6. pollinate 英 ['pɒləneɪt] 美 ['pɑːləneɪt] *vt.* 给……传授花粉
Many of the indigenous insects are needed to pollinate the local plants.
需要很多种土生昆虫给当地植物授粉。

Questions

1. What are effects of synthetic pesticides used on beneficial insects?

2. Why is it nessacery to understand the lethal effects of synthetic pesticides on beneficial insects?

3. What effect do chemical pestcides have on females?

Passage 2

The Toxicity, Persistence and Mode of Actions of Selected Botanical Pesticides in Africa Against Insect Pests in Common Beans, P. Vulgaris: A Review (Extract)

Toxicity and Persistence of Some of Active Ingredients from Botanical Pesticides

Although botanical pesticides can be used as alternatives to synthetic pesticides but the toxicity of the chemical compounds extracted from botanical pesticides to insect pests and humans, persistence to the environment and mode of actions to insect pests are not clear. Therefore, this part, intends to explain the toxicity and persistence of pyrethrin from pyrethrum, rotenone from T. vogelii, azadirachtin from Azadirachta indica, and some active chemical compounds from V. amygdalina, L. camara, and T. diversifolia. These botanical pesticides are selected because they are commonly found around our homes, along the roads, river banks and bush lands in northern part of Tanzania and may be used as botanical pesticides.

1) Rotenone

Rotenone is contained in large amount in plant species especially Tephrosia, Derris, and Lonchocarpus. All these plants are in the family fabaceae in Leguminosae. Rotenone is used as natural insecticide, piscicide, and pesticide. It is a relatively low toxicity insecticide for use in gardens but is highly toxic to fish and is sometimes used to eliminate unwanted fish from lakes. It occurs naturally in the seeds, stems, leaves and the roots of plants in fabaceae family. It is the first described member of the family of chemical compounds known as rotenoids. The LD_{50} of rotenone for rats is 132−1500 mg/kg. In human being, rotenone is moderately toxic with an oral LD_{50} ranges from 300 to 1,500 mg/kg. This compound is highly toxic to fish and insects because it is lipophilic in nature.

The respiratory mechanism of fish is directly linked to water through the gills and in insect is directly exposed through trachea whereby rotenone is easily taken up through the gills or trachea into the bloodstream of fish, and insects respectively resulting to death. However, rotenone is less toxic to mammals and birds since the route of ingestion is through the digestive tract whereby the compound is easily broken down to less toxic compounds before toxic quantities can enter the bloodstream. Rotenone is rapidly broken down by sunlight which is both an advantage and disadvantage. Since it breaks down rapidly, it does not accumulate in the environment and less harmful to non-target organisms. However, it must be reapplied at short intervals and is usually applied in the early morning or in the evening to avoid degradation of it by sunlight. In water, the rate of decomposition depends upon several factors, including temperature, pH, turbidity of water and sunlight. The half life of rotenone is four days. The half-life of rotenone in natural waters ranges from half a day at 24 ℃ to 3.5 days at 0 ℃. However, there is limited scientific information about toxicity of rotenone to organisms and persistence of it in the environment. Therefore, detail studies are needed on the toxicity of rotenone to various animals and persistence of it in the environment for using it sustainably as botanical pesticide.

2) Azadirachtin from Neem, Azadirachta Indica

Neem tree is in Meliaceae family possessing bitter triterpenoids. The active compound in the neem is azadirachtin which is found in the leaves, and also concentrated in the seeds. It is a bitter, complex chemical compound which belongs to the limonoid group and it show strong biological activities among various insect pests. This compound is a feeding deterrent and growth regulator. This compound can affect about 200 species of insects by acting as antifeedant and growth disruptor. Azadirachtin has a toxicity and fascinating effect on insects (LD_{50} (S. littoralis), 15 μg/g). It has very low toxicity to mammals whereby the LD_{50} in rats is greater than 3,540 mg/kg, which makes it practically non-toxic to mammals and also has been reported to be non-mutagenic. Azadirachtin has been found to degrade rapidly under environmental factors such as ultraviolet (UV) radiation in sunlight, heat, air moisture, acidity and enzymes present in foliar surfaces. The half-life of azadirachtin has been found to be between 48 minutes and 3.98 days under UV light and sunlight and 2.47 days on leaf surface. Therefore, there is a need to use azadirachtin as environmentally compatible insecticides, with selective toxicity to targeted pests, low toxic to plants and mammals and environmental friendly desired stability.

3) Pyrethrin from Pyrethrum, Tanacetum Cinerariifolium (Chrysanthemum Cinerariifolium)

Pyrethrum is powdered, dried flower head of the pyrethrum daisy, Tanacetum cinerariaefolium and pyrethrins active compound from pyrethrum with six related insecticidal compounds which occur naturally. There is pyrethrin I and pyrethrin II. The compounds related to pyrethrin I contain methyl group ($-CH_3$) and the compounds related to pyrethrin II contain -CO_2CH_3 group. Pyrethrins are axonic poisons and have an insect repellent effect when present in little amount. They are harmful to fish, but are less toxic to mammals and birds than many synthetic insecticides. In pure form, the rat oral LD_{50} is 1,200—1,500 mg/kg. The technical grade of pyrethrum is less toxic to rat with the LD_{50} of about 1,500 mg/kg. Pyrethrins degrade easily when exposed to the environment moisture, air and the sunlight. The half-life of pyrethrins in the environment and field-grown bell pepper fruit is two hours or less. However, there is limited scientific information about toxicity of pyrethrins compounds to various organisms and persistence of it in the environment. Therefore, detail studies are needed about the toxicity of pyrethrins compounds to various organisms and persistence of it in the environment for we can use it sustainably as botanical pesticide.

4) Sesquiterpene Lactones from T. Diversifolia

Many classes of secondary metabolites which are isolated from the T. diversifolia extracts include diterpenoids, flavonoids, sesquiterpene lactones and chlorogenic acids derivatives. However, the most abundant terpernoids in T. diversifolia are sesquiterpene lactones. But tagitinins compounds which are in sesquiterpene lactones class are the most studied. The sesquiterpene lactones and diterpenoids have biological activities and contribute to inflammatory activity. T. diversifosia is used as traditional medicine in constipation, stomach pains, indigestion, sore throat, liver pains and malaria. Researchers reported that the extracts of T. diversifolia of 10 mg/kg and 100 mg/kg administered to rats for 90 days were relatively safe with some toxicity observed at 100 mg/kg. However, the later can cause damage of the liver, the kidneys and, to a lesser extent, the heart. The liver damage observed at higher doses of aqueous extracts of T. diversifolia may result from the Chlorogenic acid, while kidney damage results from the Sesquiterpene lactones. However, no clear information about toxicity of these compounds is from T. diversifolia.

5) Pentacyclic Triterpenoids from Lantana Camara

Lantana camara is recognized to be toxic to cattle, sheep, horses, dogs and goats. The active ingredients causing toxicity of Lantana camara in grazing animals is pentacyclic triterpenoids. It is one of terpernoids which result in liver damage and photosensitivity. The toxicity of L. camara to human being is undetermined, whereby numerous studies suggest that ingestion of berries from L. camara can be toxic to humans. Researchers reported that leaf extract of L. camara had excellent repellent, moderate toxic and antifeedant activities. However, other studies have found evidence which suggests that ingestion of L. camara fruit, poses no risk to humans and is in fact edible when ripe. Studies conducted in India have found that L. camara leaves can display antimicrobial, fungicidal and insecticidal properties. L. camara has also been used in traditional herbal medicines for treating a variety of ailments, such as cancer, skin itches, leprosy, rabies, chicken pox, measles, asthma and ulcers. Lantana camara has been tested as an alternative to fumigants in stored grains. Therefore, there is a need of finding out more information about the toxicity of L. camara to the health of human being.

http://www.scirp.org/journal/PaperInformation.aspx? PaperID=63091

Vocabulary

rotenone 英 ['rəutnəun] 美 ['routənoun] n. 鱼藤酮

Azadirachta indica 印度苦楝树

deterrent 英 [dɪ'terənt] 美 [dɪ'tɜːrənt] n. 制止物；威慑物

fabaceae 英 [fə'beisiiː] 美 [fə'beisiiː] n. 豆科；蝶形花科

piscicide 英 ['pɪsɪsaɪd] 美 ['pɪsɪsaɪd] n. 灭鱼；灭鱼药

gill 英 [gɪl] 美 [gɪl] n. 菌褶；（鱼等的）鳃

trachea 英 [trə'kiːə] 美 ['treɪkiə] n. 气管；导管；螺旋纹管

ingestion 英 [ɪn'dʒestʃən] 美 [ɪn'dʒestʃən] n. 摄取；采食

decomposition 英 [ˌdiːkɒmpə'zɪʃn] 美 [ˌdikɑmpə'zɪʃən] n. 分解；腐烂

Azadirachta indicaNeem tree 印楝树

meliaceae 英 [miːˈliəsiˌiː] 美 [miːˈliəsiˌiː] n. [植] 楝科

triterpenoid 英 [traɪ'tɜːpɪnɔɪd] 美 [traɪ'tɜːpɪnɔɪd] n. 三萜系化合物

limonoid group 柠檬苦素类

mutagenic 英 [ˌmjuːtə'dʒenɪk] 美 [ˌmjuːtə'dʒenɪk] adj. 诱导有机体突变的物质

Tanacetum cinerariifolium Pyrethrum 除虫菊

daisy 英 ['deɪzi] 美 ['dezi] n. 雏菊；菊科植物

sesquiterpene lactone 倍半萜内酯

fungicidal 英 [ˌfʌndʒɪ'saɪdl] 美 [ˌfʌndʒə'saɪdəl] adj. 真菌的；由真菌引起的

rabies 英 ['reɪbiːz] 美 ['rebiz] n. [医] 狂犬病；恐水病

chicken pox 英 ['tʃikin pɒks] 美 ['tʃɪkən pɑks] n. 水痘

Useful Expressions

1. degrade 英 [dɪ'greɪd] 美 [dɪ'gred] vt. 降低；贬低；使降级；降低……身份

You should not degrade yourself by telling such a lie.

你不该说那样的谎话降低自己的人格。

2. antifeedant 英 [æntɪfiːˈdænt] 美 [æntɪfiːˈdænt] n. 拒食素（一种能使植食性昆虫产生拒食现象的化学物质）

The essential oil showed strong non-selective antifeedant and oviposition deterrence effect, but the persistence was not good.

龙柏精油对小菜蛾具有较强的非选择性拒食作用和产卵忌避作用,但由于精油的挥发效果不持久。

Questions

1. What about the safety of botanical pesticides?
2. What damages rotenone?
3. Why is it nessacery to use azadirachtin as environmentally insecticides?

8.5 Development of Herbal Feed Additive

Passage 1

Review: Chinese Herbs as Alternatives to Antibiotics in Feed for Swine and PoultryProduction: Potential and Challenges in Application (Part 1)

It is generally accepted that enteric infection and its associated gastrointestinal dysfunction is the major stress that impairs intestinal function and compromises the immunity of food-producing animals, consequently leading to growth retardation and increased morbidity and mortality of animals. Strategies, for example, the use of enzymes, organic acids, antibiotics, probiotics, prebiotics, and active phytochemicals for controlling enteric infection, improving animal immunity, and ameliorating intestinal function (including modulation of intestinal microbiota) are common approaches to combat stress and to maximize animal production. In the past several decades, antibiotics have been used extensively in the commercial production of food-producing animals to prevent enteric infection and to promote animal growth, which has contributed significantly to agricultural economy and food supply security worldwide. However, this practice has been under close scrutiny because it threatens public health due to antibiotic residues in food animal products and the spread of antibiotic resistance in zoonotic bacterial pathogens. The latter erodes the therapeutic effectiveness of clinically important antibiotics in human medicine. To address public concerns, the European Union (EU) passed legislation in 2006 to ban the use of antibiotics as animal growth promoters. More countries are expected to follow. In fact, the United States Food and Drug Administration (FDA) announced a regulation to promote the judicious use of medically important antibiotics in food-producing animals. The agency sets a 3-year timetable for phasing out the use of antibiotics to enhance growth of food-producing animals, and currently asks the livestock and drug industries to voluntarily cut their use. In Canada, the Advisory Committee on Animal Uses of Antimicrobials and Impact on Resistance and Human Health established by Health Canada proposed the prudent use of antibiotics in 2002, including making all antibiotics used for disease treatment and control available by prescription only. Therefore, development of viable alternatives to antibiotics in feed is becoming an urgent need and is currently under active investigation worldwide. In general, a viable alternative to dietary antibiotics is expected to have the capacity to effectively control enteric infection, improve animal immunity, or ameliorate intestinal function. Additionally, it should meet the following criteria: (i) not compromising animal production, (ii) not threatening human and public health, (iii) being economically viable, and (iv) being environmentally sound. These requirements clearly indicate the complexity and challenges in developing viable alternatives, which also explains why there is currently no single alternative that is fully capable of being a substitute for dietary antibiotics.

Traditional Chinese medicine has developed a unique theoretical framework and abundant clinical experience through a long history of clinical practice to ameliorate human and animal health. It shares the same basal philosophical theory as of the "holistic concept". With this basal concept, the clinic practice relies on syndrome-based treatments and multi-targeting/multi-component modes of action, which ensures its irreplaceable advantages in the treatment of complex diseases or disorders. The first use of Chinese herbs can be traced back to over 2,000 year ago in China to promote animal health and production. Recent studies on Chinese herbs as feed supplements have also shown that they can modulate nutritional metabolism, immune responses, and the intestinal health of food-producing animals. The Chinese herbs thus provide an attractive alternative to dietary antibiotics for the commercial production of food-producing animals. Among the Chinese herbs or their components tested thus far, some traditional Chinese herbal formulas (TCHFs) and polysaccharides have demonstrated the most promise. Their effect on promoting animal growth performance is comparable to that of dietary antibiotics. Some commercial products, such as Multifunctional Chinese Herbal Additive, have been developed from these formulated herbs and herbal polysaccharides and are widely used in the commercial production of swine and poultry in China. However, due to cultural and philosophical differences, traditional Chinese medicine theory has been considered abstract, complex, and difficult to understand by western medical professionals and is thus hard to be recognized by western medical communities. In addition, the paucity of well-controlled clinical trials and occasional manifestations of toxicity have impeded the acceptance of traditional Chinese medicine into mainstream of western medicine. In recent years, with the advent of top-down systems biology theory and tools of whole-body systems biology (WBSB), there is a new opportunity to study traditional Chinese medicine and to close the gap between traditional Chinese medicine and western medicine, because of the similarity in the theoretical foundations between traditional Chinese medicine and WBSB. Nonetheless, there are some issues to be addressed before traditional Chinese medicine can reach its full potential. Typically, these include insufficient source supplies of traditional Chinese medicine and quality control in production and processing as well as the lack of scientific evaluation standards/systems and a full understanding of the mechanisms underlying the functions. In addition, regulatory approvals for the use of traditional Chinese medicine could be a challenge, particularly with the use of a TCHF when multi-components are involved and a clear cause-effect relationship has not been established yet. This article critically reviews recent progresses in scientific research of Chinese herbs as feed additives for promoting swine and poultry production. We have focused on the potential of Chinese herbs as an alternative to dietary antibiotics and possible challenges in their future application. In addition, we have taken advantage of our bilingual skills in both English and Chinese and have carefully used scientific information from well-recognized Chinese journals, which can be regarded as a unique feature of this review article.

http://www.nrcresearchpress.com/doi/10.4141/cjas2013-144

Vocabulary

enteric 英 [en'terɪk] 美 [ɛn'tɛrɪk] *adj.* 肠的
dysfunction 英 [dɪs'fʌŋkʃn] 美 [dɪs'fʌŋkʃən] *n.* 机能障碍；机能失调
morbidity 英 [mɔː'bɪdətɪ] 美 [mɔː'bɪdətɪ] *n.* 发病率；发病；病态
scrutiny 英 ['skruːtəni] 美 ['skrutni] *n.* 监督；细看；细阅
zoonotic 英 [,zəʊə'nəʊtɪk] 美 [,zoʊr'noʊtɪk] *n.* 动物传染病的
ameliorate 英 [ə'miːliəreɪt] 美 [ə'miljə,ret] *vt.* 改良；减轻(痛苦等)

Useful Expressions

1. impair 英 [ɪm'peə(r)] 美 [ɪm'per] *vt.* 损害；削弱
Consumption of alcohol impairs your ability to drive a car or operate machinery.
饮酒会削弱你驾驶汽车或操控机器的能力。

Questions

1. What erodes the therapeutic effectiveness of clinically important antibiotics in human medicine?

2. What do the westerners think of traditional Chinese medicine theory?

3. What problems should be dealt with to reach traditional Chinese medicine' full potential?

Passage 2

Chinese Herbs as Alternatives to Antibiotics in Feed for Swine and Poultry Production: Potential and Challenges in Application (Part 2)

Alternatives to Antibiotic Growth Promoters

Enteric infection and its associated gastrointestinal dysfunction cause major stress to food-producing animals, particularly to young ones, which can impair intestinal metabolism and functions, and compromise the immunity. Consequently, growth in animal production is retarded, and morbidity and mortality are increased. Dietary antibiotics work primarily in the digestive tract of animals. One of the major advantages of using antibiotics is their effectiveness in controlling enteric bacterial pathogens in addition to functions in the modulation of the host immune response and intestinal microbiota composition and richness. Therefore, the promotion of animal intestinal health is expected to be the primary function of a viable alternative. To achieve this promotion, different strategies and approaches can be used, such as reducing intestinal pH value, maintaining protective intestinal mucins, selection for beneficial intestinal microbes and exclusion of pathogens, enhancing fermentation and short-chain fatty acid production, improving nutrient uptake, and increasing the humoral immune response.

Similarity Between WBSB and Traditional Chinese Medicine

Recently, Nicholson proposed the theory of WBSB in 2006. The WBSB regards the host (human or animal) body as a complex "super-organism", consisting of cells of the host and symbiotic microorganisms, and adopts a "top-down" strategy to reflect the emergent functions of the super-organism in a holistic context. This theory is similar to the "holistic concept" of traditional Chinese medicine. In clinical practice, the major methods used by traditional Chinese physicians to diagnose and treat diseases include inspection, auscultation and olfaction, interrogation, and pulse-taking and palpation. Eight principles, namely, yin-yang, exterior-interior, cold-heat, and deficiency-excess, for syndrome differentiation are also under consideration for diagnoses, based on the traditional Chinese medicine theory and clinical experiences. A traditional Chinese veterinary physician prescribes appropriate medicines, normally a traditional Chinese medicine formula, to sick animals based on their empirical identification of the traditional Chinese medicine syndromes. The traditional Chinese medicine formula is composed of the main function in the monarch drugs, minister drugs, assistant drugs, and guide drugs, according to the compatibility theory and dialectical view. The monarch drugs are the key ingredient in the prescription; the minister drugs promote the monarch drugs to exert curative effect; the assistant drugs strengthen the effect of the prescription or restrict toxins; and the guide drugs direct other ingredients to work on the affected part (or organs) in the host. However, this empirical diagnostic and treatment in traditional Chinese medicine is difficult to be understood and recognized by western veterinarians. The WBSB conducts measurements at the whole-body level of the host. There are at least three types of samples containing information at the whole-body level, including blood, urine, and stool for the human or animal body, similar to traditional Chinese medicine, but with high-throughput, high-precision, and molecular-marker-based instrumental analyses, in addition to undertaking differentiation using multivariate statistical methods. In this regard, the WBSB would help bridge the gap between traditional Chinese medicine and western medicine and interpret traditional Chinese medicine through evidence-based methods, which could eventually lead to the merger of traditional Chinese medicine and western medicine into a new medical system.

Rwcent Mechanistic Studies on Traditional Chinese Medicine with the "-Omics" Techniques

In recent years, the development of information-rich "-omics" techniques, for example, genomics, proteomics and transcriptomics, has provided an opportunity to simultaneously analyze a large number of genes/targets associated with dietary interventions in humans and animals. Such technologies have also been applied to determine the multi-target action of traditional Chinese medicine in humans and animals. An interesting example is the current studies on a model traditional Chinese medicine herbal formula, Si-Wu-Tang [Si-Wu decoction (Chinese name), Samultang (Korean name), or Shimotsu-to (Japanese name)], one of the most popular traditional oriental medicines for women's health. The major function of Si-Wu-Tang is to improve a deficiency of Qi and Blood. The Si-Wu-Tang has also shown sedative, anti-coagulant and anti-bacterial activities, as well as a protective effect on radiation-induced bone marrow damage in model animals, although the mechanisms and active constituents are unknown. With microarray and transcriptome analysis, the gene expression profile of differentially expresses genes related to Si-Wu-Tang intervention consistent with Si-Wu-Tang's widely claimed use for women's diseases and indicates a phytoestrogenic effect, therefore clearly explores the mechanisms of actions for therapies in humans. In addition, high-throughput, information-rich assays can also be used to fingerprint herbs and their active extracts. Many studies have thus begun to focus on the quality control and sample variability of traditional Chinese medicine, which is pivotal for reproducibility and standardization of biological effects in the study of herbal preparations. Such applications of information-rich approaches make the use of "-omics" techniques particularly appropriate for addressing many of the issues encountered in traditional Chinese medicine research, which have hampered its acceptance in western medicine, and thus more integration with western medical practice.

General Properties of Chinese Herbal Feed Aditives

Over thousands of years of practice and experience, practitioners of traditional Chinese medicine have been able to diagnose, cure, and prevent many diseases through the administration of complex mixtures derived largely from plants, mineral and animal products. Undoubtedly, this approach has worked at least within the social and economic environment in China. Chinese herbal feed additives (CHFAs) are commonly defined as crude Chinese herbs or their-derived compounds which are dietary supplements at sub-clinical doses to improve the productivity of food-producing animals through the amelioration of their immune status, promotion of growth performance, or improvement of the quality of food products from the animals. The CHFAs have become widely used in the commercial production of swine and poultry in China. These additives can be further classified with respect to their origin and processing, such as herbs (single or herbal formula, from seeds, flowers, leaves, stems or roots of herbal plants), herbal crude extracts (aqueous or methanol dissolved phytogenic compounds, extracted from a single or a recipe composed of two or more herbs), herbal polysaccharides (soluble or insoluble polysaccharides derived from herbs), etc. Although Chinese herbs have a 1,000-year history of use in animal husbandry in China, their use as a growth promoter in the large-scale commercial production of food animals had not received any significant research attention until the recent legislation was issued to restrict the use of dietary antibiotics in food-producing animals in the EU countries. It should be emphasized that the normal practice in using CHFAs for animal care and production in China is also guided by the empirical diagnostic and treatment knowledge of traditional Chinese medicine. In general, the additives are supplemented at different sub-clinical doses to feed healthy animals for both nutritional and health care purposes, depending on the growth stage during the entire production period. Additionally, the safety and efficiency of Chinese herbs have been well documented throughout the long practice of traditional Chinese medicine. The CHFAs can provide animals not only with nutrients, such as protein, small peptides, essential amino acids, oligosaccharides, fatty acids, starch, vitamins and organic trace minerals, but also with many bioactive ingredients with anti-microbial activities, immune enhancement, and stress-reduction properties. The primary mode of action through which CHFAs promote animal growth includes decreasing stress in young animals and the promotion of their intestinal health and functions. In particular, the establishment of a healthy gastrointestinal ecosystem with a healthy microbiota is critical for controlling potential pathogens and digestive disorders in the critical phases of the life cycle of young animals, such as the weaning of piglets and the early days of newly hatched chicks.

http://www.nrcresearchpress.com/doi/10.4141/cjas2013-144

Vocabulary

mucin 英 ['mju:sɪn] 美 ['mju:sɪn] *n.* 黏液素
auscultation 英 [,ɔːskəl'teɪʃn] 美 [,ɔskəl'teʃən] *n.* 听诊
olfaction 英 [ɒl'fækʃn] 美 [ɑl'fækʃən, ol-] *n.* 嗅；嗅觉
interrogation 英 [ɪn,terə'geɪʃn] 美 [ɪn,terə'geɪʃn] *n.* 讯问；审问；疑问句
palpation 英 [pæl'peɪʃn] 美 [pæl'peɪʃn] *n.* 触诊；扪诊
proteomics 英 [p'rəʊtiːəʊmɪks] 美 [p'roʊtiːoʊmɪks] *n.* 蛋白质组学
transcriptomics 英 [trænsk'rɪptəʊmɪks] 美 [trænsk'rɪptoʊmɪks] *n.* 转录物组学
starch 英 [stɑːtʃ] 美 [stɑːrtʃ] *n.* 淀粉；含淀粉的食物
monarch drug 君药；主药
minister drug 臣药
assistant drug 佐药
guide drug 导向药物

Useful Expressions

1. retard 英 [rɪ'tɑːd] 美 [rɪ'tɑːrd] *vt.* 使减速；妨碍；阻止；推迟
Continuing violence will retard negotiations over the country's future.
持续不断的暴力活动会阻碍有关国家未来的谈判的进行。
2. husbandry 英 ['hʌzbəndri] 美 ['hʌzbəndri] *n.* 农业；资源管理；妥善管理
They depended on animal husbandry for their livelihood.
他们以畜牧业为生。
3. additive 英 ['ædətɪv] 美 ['ædɪtɪv] *n.* 添加剂；添加物；[数]加法
Strict safety tests are carried out on food additives.
对食品添加剂进行了严格的安全检测。
4. wean 英 [wiːn] 美 [win] *vt.* 使断奶；使断念
The baby would be weaned and she would bring it home.
婴儿断奶后，她会把婴儿带回家。

Questions

1. What causes pressure to food-producing animals?
2. Why is the promotion of animal intestinal health expected to be the primary function of a viable alternative?
3. What are Chinese herbal feed additives?

8.6 Development of New Medicinal Plants Resources

Passage 1

National Medicinal Plants Board for Development of Medicinal Plants Sector

The government of India set up the National Medicinal Plants Board (NMPB) vide resolution notified on 24th November, 2000 to co-ordinate with ministries/departments/organizations/state/UT governments for development of medicinal plants sector in general and specifically in the areas relating to assessment of demand supply, advising on policy, promotion of conservation, proper harvesting, cultivation, quality control, research and development, processing, marketing of raw material in order to protect, sustain and develop this sector.

For overall development of medicinal plants sector in the country the NMPB has been implementing following schemes since year 2008—2009:

(i) Central Sector Scheme for "Conservation, Development and Sustainable Management of Medicinal Plants" aimed at providing support for survey, inventorization, *in-situ* conservation, *ex-situ* conservation/herbal gardens, research and development, linkage with peoples collectives like Self Help Groups (SHGs), Joint Forests Management Committees (JFMCs), etc. The scheme is being continued during the 12th Plan.

(ii) Centrally Sponsored Scheme of "National Mission on Medicinal Plants" is primarily aimed at supporting cultivation of medicinal plants on private land with backwards linkages, for establishment of nurseries for supply of quality planting material, etc., and forward linkages for post-harvest management, marketing infrastructure, certification, etc. Currently this scheme is being implemented as a component (Medicinal Plants) of the National AYUSH Mission (NAM) Scheme of the Ministry of Ayurveda, Yoga and Naturopathy, Unani, Siddha and Homoeopathy (AYUSH).

For increasing the availability of medicinal plants in the country, the NMPB under its aforesaid schemes is mainly supporting the activities/programmes for augmenting the existing resources and cultivation of prioritized medicinal/herbal plants species.

Presently, as such there is no region specific programme/scheme for the cultivation of herbal plants which are endemic to specific region. However, the programme of cultivation of medicinal/herbal plants is being implemented in the states under the Scheme of NAM. The scheme is being implemented in a mission mode through mission director identified in the states. The state mission directors prepare the state annual action plans (SAAPs) for cultivation of prioritized medicinal/herbal plants which are being approved by the directorate of the NAM. The NMPB supports cultivation of herbal plants by providing financial assistance in the form of subsidy at 30%, 50% and 75%. The endemic and endangered herbal plants are generally kept at category eligible for 75% subsidy, so as to promote their cultivation in a large scale.

This information was given by the minister of state (independent charge) of the Ministry of AYUSH, Shri Shripad Yesso Naik in reply to an unstarred question in Lok Sabha today.

http://www.newstrackindia.com/newsdetails/2015/8/1/1-National-Medicinal-Plants-Board-for-Development-of-Medicinal-Plants-Sector.html

Vocabulary

vide 英 ['viːdeɪ] 美 ['vaɪdi, 'viˌde, 'wi-] *v.* 请见；参阅
resolution 英 [ˌrezəˈluːʃn] 美 [ˌrezəˈluʃən] *n.* 分辨率；解决；决心；坚决
aforesaid 英 [əˈfɔːsed] 美 [əˈfɔrˌsɛd, əˈfor-] *adj.* 上述的；前述的（常用于法律文件）
annexed 英 [əˈnekst] 美 [əˈnekst] *adj.* [法] 附加的；附属的
unstarred 英 [ʌnˈstɑːd] 美 [ʌnˈstɑːd] *adj.* [计] 未加星号的

Useful Expressions

subsidy 英 ['sʌbsədi] 美 ['sʌbsɪdi] *n.* 补贴；津贴；助学金；奖金
European farmers are planning a massive demonstration against farm subsidy cuts.
欧洲农场主正策划一场抗议削减农场补贴的大规模示威游行。

Questions

1. Why has the Government of India set up the National Medicinal Plants Board?
2. What has the NMPB been doing since year 2008?
3. What has the NMPB been doing to increase the availability of medicinal plants in the country?

Passage 2

Development of Medicinal Plant Gardens in Aburi, Ghana

Botanic Gardens Conservation International was fortunate this year to win a U.K. National Lotteries Charities Board grant of £79,900 to support a new project in Ghana. This project is undertaken in collaboration with the Aburi Botanical Gardens in Eastern Ghana to promote the cultivation of medicinal plants in home gardens.

The greatest threats to the remaining forests in Ghana are agricultural encroachment and fire in drought years. Two of Ghana's national parks have been treated as a centre of plant diversity site and probably represent the only area of Ghana with relatively undisturbed rain forest. As most plant species used in primary health care are collected from the wild, habitat destruction is seriously affecting their availability, as is collection pressure with severe strain being put on plant populations in the vicinity of urban centres. There is consequently an urgent need to encourage local people to cultivate medicinal plants for use in their community.

Sustainable Futures

The use of medicinal plant species and the accumulated knowledge of traditional medicinal practice are threatened by habitat destruction and by the unsustainable harvesting of plants from the wild. The resulting shortage of plant materials has been noted by collectors concerned at having to travel further for raw materials. This is a valuable indicator of the current status of medicinal plant species in the wild and is a critical warning sign that action needs to be taken now, to reduce pressure on these diminishing populations.

For several years, the Aburi Botanical Gardens have received requests for medicinal plants from the garden as well as for seed and information on the cultivation of medicinal plant species. Religious groups have requested medicinal plant material from the garden, for use in rituals and religious instruction, and schools regularly request information and samples of medicinal plant material for education.

The knowledge of herbal medicine is extensive and varies from one region of the country to another. It is hoped that this project will draw together a number of traditional healers to contribute their knowledge and experience. Their skills at identifying species and at monitoring their availability in the wild will be an invaluable asset to the project.

Plant conservation and sustainable utilisation are one and the same thing and it is critical that local people are involved and empowered to develop sustainable practices for growing and harvesting medicinal plants.

Aims

By encouraging villagers to cultivate and trade in medicinal plants, the project aims to increase access to preventative and primary health care. By promoting the sustainable use and cultivation of native medicinal plants, the project hopes to provide locally accessible and plentiful supplies of raw materials and a means to develop small-scale commercial production. The project aims to reduce the harvesting pressure currently being exerted on wild native medicinal plants and directly benefit the rural people who can be assured that local natural resources will be available for future generations.

This project aims to complement on-going work in species conservation in Ghana and will reinforce a Darwin Initiative funded project of Botanic Gardens Conservation International (BGCI) and the Aburi Botanical Gardens in partnership with the Legon University Herbarium, the Center for Remote Sensing and Geographic Information

Services (CERGIS) at Legon University, the World Conservation Monitoring Centre (WCMC) in Cambridge, U.K. and the Royal Botanic Garden Edinburgh (RBGE), U.K.

The Project Team

Mr. Theodoplius Agbovie, the current curator of the Gardens, has been appointed the full-time project coordinator in Ghana, and he is assisted by Miss Linda Afriyie Damankah who has already begun the compilation of a Ghanaian medicinal plant list. This has been used as a primary working list for the project's ethnobotanical survey, now underway. William Ofusu Hene, recently retiring from the Medicinal Plant Research Centre at Mampong (about 8 km from the Gardens) has also joined the survey team. His expertise will greatly benefit the identification of traditional herbal preparations and the harvesting methods commonly used. The U.K. project coordinator, based in London, is Fiona Dennis of BGCI.

A project management committee has been set up from which a project advisory group can be identified. This group will consist of representatives from a number of different organisations and user groups such as traditional healers, church groups, local schools and local women's groups in Eastern Ghana.

Medicinal Plant Garden

The project coordinator (Ghana), working with garden staff and the project advisory group, has started work on a model medicinal plant garden at the Aburi Botanical Gardens. The land (136 acres) has been generously donated by Aburi Botanical Gardens within the gardens for the creation of a model garden. This model garden will be used to encourage local people and schools to set up their own medicinal plants and herb gardens.

The allotted land consists of mature trees and under-storey shrubs. This forest has suffered from the encroachment of local farmers in search of fuel wood, timber and medicinal plants and the accidental introduction of a number of alien palm and bamboo species. However, it is clear that the majority of target species for this project will be tree species and the existing mature trees within this plot are of great value.

The model garden will use the existing mature trees for the overall structure of the garden and will increase the species diversity by under planting with species grown in the Gardens' nursery. Each significant plant will be identified and labelled. Interpretation plaques will be erected at appropriate sites throughout the garden.

The next step will be the plant diversity survey to identify exactly what the existing flora can contribute to the biodiversity of the garden. Following this activity the enrichment of this plant community will begin.

There will also be an extensive education programme in the garden. The project coordinators will work with local people to construct signs that inform visitors about the value of medicinal plants, the threat to them from over-collection and the conservation aims of the project.

The project will identify the species of plants most commonly used by local people in medicine. This will be carried out in collaboration with members of the project advisory committee. A target list of plants for cultivation will be drawn up based upon their use and estimated availability in recent years. The project will establish protocols for propagation and cultivation to ensure that wild species can be successfully brought into home cultivation.

Fiona Dennis (BGCI) visited the gardens in October 1999 and met the nursery staff of the project. A *pro forma* collecting sheet and nursery records book, including details on propagation methods and nursery conditions were drawn-up. This will ensure that the recording and monitoring of all species coming into, and being planted out into the model garden will be fully documented. Detailed records will enable a thorough analysis of the procedures required to successfully propagate and cultivate these species. This information will provide the basis upon which to write the training booklet.

Training Booklet and Workshops

In collaboration with the project management committee, the project team will produce a training booklet. This booklet will illustrate the cultivation, propagation, harvesting, drying and storing of medicinal plants.

This booklet will be used in the workshops and practicals which will be run as parts of the project. These workshops will help local schools and communities set up their own gardens. The project will also provide seeds, polythene bags and fertilisers to help stimulate interest in the cultivation of medicinal plants.

Community Gardens

The survey team has visited several villages and spoken to a number of traditional healers. From these visits they have been able to identify species, harvesting methods and traditional medicinal plant uses.

Three villages have already agreed to work within the project and establish co-operatives for the cultivation of medicinal plant species, both for use within the village and for small-scale commercial production. It is hoped that success with these villages will help encourage surrounding villages to make a commitment to cultivation within the duration of this project.

Local Trade

By working with local people to develop best practices for the harvesting, drying and storage of medicinal plants, the project hopes to stimulate local trade both within and outside the town of Aburi, and to help increase people's economic independence. The gardens will also provide a clearing-house and central depository for large-scale harvests and will work closely with the Mampong Centre for Research into Traditional Medicine. This centre, staffed by western trained doctors carries out research into the uses, potency and commercial value of herbs and plants. The research centre is able to utilise a great deal of plant materials and are keen to develop sustainable supply routes for many of their manufactured plant products. As part of the Ministry of Health, the research centre is also able to ensure that the plant medicines manufactured there, meet regulatory standards.

The project will aim to change attitudes and present an opportunity for responsible conservation at a local level. Not all species of medicinal plants identified during the project will be adaptable to cultivation, particularly tree species, however, the presentation of the conservation issues involved in destructive wild harvesting will highlight the need not only for cultivation, but for sustainable wild harvesting techniques.

http://www.bgci.org/resources/article/0063/

Vocabulary

grant 英 [grɑːnt] 美 [grænt] *n.* 拨款；补助金
encroachment 英 [ɪnˈkrəʊtʃmənt] 美 [ɪnˈkrotʃmənt] *n.* 侵入；侵占；侵蚀
vicinity 英 [vəˈsɪnəti] 美 [vɪˈsɪnɪti] *n.* 附近地区；附近；邻近
preventative 英 [prɪˈventətɪv] 美 [prɪˈvɛntətɪv] *adj.* 预防性的
allotted land　划拨土地
under-storey　下层林木
protocol 英 [ˈprəʊtəkɒl] 美 [ˈproʊtəkɔːl] *n.* （数据传递的）协议；科学实验报告
propagation 英 [ˌprɒpəˈgeɪʃn] 美 [ˌprɑpəˈgeʃən] *n.* 宣传；传播；[生]繁殖法
pro forma 英 [ˌprəʊˈfɔːmə] 美 [ˌproʊˈfɔːrmə] *adj.* 形式上的；预计的
training booklet　培训手册
potency 英 [ˈpəʊtnsi] 美 [ˈpoʊtnsi] *n.* 效力；潜能；权势
plaque 英 [plæk] 美 [plæk] *n.* 匾；饰板

Useful Expressions

1. in collaboration with　与……合作
Scientists hope the work done in collaboration with other researchers may be duplicated elsewhere.
科学家们希望与其他研究人员合作的成果可以适用于别处。
2. availability 英 [əˌveɪləˈbɪləti] 美 [əˌveləˈbɪlɪti] *n.* 有效；可利用性；可得到的东西（或人）
The nature and availability of material evidence was not to be discussed.
实质性证据的性质和可用性毋庸讨论。

Questions

1. What are the greatest threats to the remaining forests in Ghana?
2. What threats the use of medicinal plant species and the accumulated knowledge of traditional medicinal practice?
3. What are the aims of the project mentioned in this passage?
4. What should be done in the project?
5. Who has produced the booklet and what does it contain?

Chapter Nine

Investigation and Study on Medicinal Plant Resources

9.1 Purpose and Significance of Investigation and Study on Medicinal Plant Resources

Passage 1

MaxEnt Modeling for Predicting the Potential Distribution of Endangered Medicinal Plant (H. Riparia Lour) in Yunnan, China (Part 1)

Introduction

An organism's habitat is the combination of the space it inhabits and all eco-factors in that space, including the abiotic environment and other organisms that are necessary for the existence of individuals or groups. Habitat quantity and quality have a significant impact on a species' distribution and richness within environment. Habitat loss affects the spatial pattern of residual habitat and induces microclimatic change and habitat fragmentation. Thus, habitat loss has negative effects on species richness that may be of long duration and high intensity. Habitat loss is the main reason for species endangerment, species extinction and biodiversity loss. Several reasons, such as climate change and land use change, may shrink, degrade or destroy the habitats of wild animals and plants. Ilex khasiana Purk, a tree species of northeastern India, was critically endangered by habitat loss; only approximately 3,000 individuals of Ilex khasiana Purk currently survives. The demands of an ever-increasing human population—the most important being of land for agriculture, industry and urbanization—have strong impacts on the habitat of Malabar nut (Justiciaadhatoda L.), a medicinal plant. The population of Malabar nut is shrinking in India's Dun Valley due to habitat loss. By 2010, approximately one-fifth of all of the world's plants species were at risk of extinction.

H. riparia Lour is a rheophyte native to Yunnan Province. It is a medicinal plant with high ecological and economic values. Its abundance has decreased sharply in recent decades. A field investigation in 1984 (before the construction of Manwan reservoir) showed that H. riparia Lour was present in at least four habitats in the Manwan reservoir area, and its abundance was significantly more than 400; only one habitat among these four remained in 1997 (after Manwan reservoir's construction). Two habitats in the

Manwan reservoir lake and one habitat below Manwan dam were flooded. The only remaining H. riparia Lour is scattered throughout the floodplain between the upstream stretches and the estuary of Luozha river, but their condition in 1997 was worse than that in 1984. However, few studies of the habitat quality of H. riparia Lour have been undertaken. Consequently, researching the habitat preferences of H. riparia Lour, developing a habitat suitability model to calculate the spatial distribution of this species, and seeking suitable survival conditions for H. riparia Lour are crucial to its conservation.

The first task was to understand how the environment structures the distribution of H. riparia Lour. To do so, we built a species distribution model (SDM) as a function of climate, topography and location. SDM mainly uses distribution data of species (presence or absence) and environmental data to algorithmically estimate species' niches, and then project those niches onto the landscape, reflecting a species' habitat preferences in the form of a probability. The results can be explained as the probability of species presence, species richness, habitat suitability, and so on. SDM has been used to predict the ranges of plant diseases and insects, model the distributions of species, communities or ecosystems, assess the impact of climate, land use and other environmental changes on species distributions, evaluate the risk of species invasion and proliferation, identify unsurveyed areas with high suitability for precious endangered species, contribute to the site selection of natural preserves, and identify target areas for species reserves and reintroductions. Typical SDM includes MaxEnt, BIOCLIM, DOMAIN, GAM, GLM, BIOMAPPER, and so on.

SDM is based on presence and absence data, which may be obtained from field investigation, specimen records, and literatures. In practice, it is very difficult to obtain absence data. Even when absence data can be obtained, it is unreliable. Presence data for rare and endangered species is also limited. Elith et al. used 16 methods to model the distributions of 226 species from six regions around the globe. The results indicated that the predictive ability of Maxent was always stable and reliable, and it outperformed DOMAIN, BIOCLIM, GAM, and GLM, etc., for presence-only data. As a result, among SDM, MaxEnt was selected because of its various advantages: (i) The input species data can be presence-only data; (ii) both continuous and categorical data can be used as input variables; (iii) its prediction accuracy is always stable and reliable, even with incomplete data, small sample sizes and gaps; (iv) a spatially explicit habitat suitability map can be directly produced; and (v) the importance of individual environmental variables can be evaluated using a built-in jackknife test.

This study used occurrence records of H. riparia Lour to model its habitat suitability distribution to better conserve its habitat using the following approach: (i) key environmental variables highly correlated with H. riparia Lour's distribution were selected; (ii) a MaxEnt model was developed to quantify the relationship between H.

riparia Lour's presence and the selected environmental variables (including position, topographical variables and bioclimatic variables); (iii) the habitat suitability of H. riparia Lour in three historical periods (1950—1959,1975—1985 and 2000—2009) were simulated using the developed model; and (iv) the potential habitats of H. riparia Lour under two climate warming scenarios (RCP2.6 and RCP8.5, given by the Intergovernmental Panel on Climate Change) were predicted.

Study Area and Species

1) Study Area

Yunnan Province is located in southwestern China, (21°8′N—29°15′N, 97°31′E—106°11′E). With a total area of approximately 390,000 km^2. The north side is higher than the south side in Yunnan Province, and significant temperature differences exist between the north and south. The climate of Yunnan varies regionally and with altitude. The seasonal temperature difference is small, and the diurnal temperature difference is large. Rainfall is plentiful, with clearly delineated wet and dry seasons, but precipitation is not uniform throughout the province. The annual precipitation in most parts of Yunnan is approximately 1,100 mm, but precipitation in the southern part may reach 1,600 mm. Thus, the seasonal distribution and regional distribution of precipitation are uneven; rainfall in the winter is sparse, and in the summer is abundant.

The special geographical location and complex natural environment of Yunnan Province have is the richest in wildlife species and ecosystem types. There are many endemic genera, endemic species, and rare and endangered species. Over half of all of China's species can be found in Yunnan Province, and 67.5% of all of the rare species in the country are found in Yunnan, ranking first in China. Due to natural and geographical restrictions and other factors, the diversity of biological resources in Yunnan is vulnerable. Yunnan is one of the 17 key regions of biodiversity in China and one of the 34 global biodiversity hotspots. Its biodiversity ranks first in China and attracts worldwide attention.

2) Species

H. riparia Lour (Euphorbiaceae) is an evergreen shrub 1—3 m high. H. riparia Lour has many aliases in China, such as Shuima, Xiagongchashu, Shuizhuimu, and Xiyangliu. It is an endangered species native to Yunnan Province, China. In China, it is mainly found in Yunnan Province, but it is also distributed in Guangxi Zhuang Autonomus Region and Guangdong Province. It is a medicinal plant with high ecological and economic value. Its root can be used to treat hepatitis, joint pain, stomachache, or empyrosis, and it has detoxifying and diuretic effects.

H. riparia Lour is a rheophyte commonly found along the banks of drains and on rocky flood plain, but it also grows in shrub and on riverine and in regularly flooded areas. H. riparia Lour is a deep-rooted tree; it has good resistance to drowning and scouring, and effectively prevents erosion, fixes sands and reinforces dikes. In recent decades, its population in Yunnan Province has declined significantly.

http://www.sciencedirect.com/science/article/pii/S0925857416302245

Vocabulary

abiotic 英 [ˌeɪbaɪˈɒtɪk] 美 [ˌeɪbaɪˈɑːtɪk] *adj.* 无生命的；非生物的

microclimatic 英 [maɪkrəukˈlaɪmætɪk] 美 [maɪkroukˈlaɪmætɪk] *adj.* 小范围气候的（microclimate 的变形）

upstream 英 [ˌʌpˈstriːm] 美 [ˌʌpˈstrim] *adj.* 逆流而上的；向上游的

estuary 英 [ˈestʃuəri] 美 [ˈeri] *n.* 港湾；（江河入海的）河口；河口湾

topography 英 [təˈpɒɡrəfi] 美 [təˈpɑːɡrəfi] *n.* 地貌；地形学；地形测量学

niche 英 [nɪtʃ] 美 [niːʃ] *n.* 壁龛；合适的位置（工作等）；有利可图的缺口；商机

built-in jackknife test　内置刀切试验

diurnal 英 [daɪˈɜːnl] 美 [daɪˈɜːrnl] *adj.* 白天的；每日的；占用一天的；只活一天的

sparse 英 [spɑːs] 美 [spɑːrs] *adj.* 稀疏的；稀少的

endemic genera　特有属

empyrosis 英 [empaɪˈrəʊsɪs] 美 [empaɪˈroʊsɪs] *n.* 烧伤；烫伤

riverine 英 [ˈrɪvəraɪn] 美 [ˈrɪvəˌraɪn, -ˌrin] *adj.* 河的；河流的；河边的

Useful Expressions

1. outperform 英 [ˌaʊtpəˈfɔːm] 美 [ˌaʊtpərˈfɔːrm] *vt.* 做得比……更好；胜过

In recent years, the Austrian economy has outperformed most other industrial economies.

近几年，奥地利的经济发展超过了其他多数工业国。

2. delineate 英 [dɪˈlɪnieɪt] 美 [dɪˈlɪniˌet] *vt.* 勾画；描述

Biography must to some extent delineate characters.

传记必须在一定程度上描绘人物。

Questions

1. What greatly influences species' distribution and species richness within environmen?
2. What causes the change of spatial pattern of residual habitat?
3. What species are involves in the study?

Passage 2

MaxEnt Modeling for Predicting the Potential Distribution of Endangered Medicinal Plant (H. Riparia Lour) in Yunnan, China (Part 2)

Materials and Methods

1) Data Sources

Fifty-one occurrence records of H. riparia Lour in Yunnan Province were collected from databases, including field survey data in June and December 2010, the Global Biodiversity Information Facility, the National Specimen Information Infrastructure of China, the Chinese Virtual Herbarium and in the literature.

Bioclimatic variables are very biologically meaningful for defining the environmental niche of a species. Data for 19 bioclimatic variables were downloaded from http://www.worldclim.org. A geographical base map of China was obtained from National Fundamental Geographic Information System. Using the longitude and latitude coordinates and the digital elevation model (DEM), variables including the distance to the nearest river, altitude, aspect and slope, were calculated using ArcGIS 9.3. The DEM data were obtained from http://www.gscloud.cn/. All environmental data used in this model were at 30 arc-second spatial resolution (often referred to as 1-km spatial resolution).

The climate data for three historical periods (1950−1959, 1975−1985, and 2000−2009) were obtained from China Meteorological Data Sharing Service System. In the fifth IPCC report, using the total radiative forcing (RF) in 2100 as an index, four representative concentration pathways (RCPs) were set, representing scenarios in which the total RF in 2,100 will reach 2.6 W/m^2, 4.5 W/m^2, 6.0 W/m^2 and 8.5 W/m^2 over the value in 1,750 (The fifth IPCC report). Here, two scenarios, RCP2.6 and RCP8.5, were selected. Of the four RCPs, RCP2.6 is the only scenario in which global warming in 2,100 will not exceed 2℃ compared to 1850−1900. H. riparia Lour's habitat suitability distributions in each of these two scenarios were modeled. The climate data are climate projections for the years 2061 through 2080 from global climate models (GCMs) for RCP2.6 and RCP8.5. These are available at http://www.worldclim.org.

2) Variables Selection

To select variables that can contribute more predictive power to the model, eliminate multiple linearity between variables, and establish a model that has better performance with fewer variables, cross-correlations (Pearson correlation coefficient, r) and principal component analysis (PCA) of 19 bioclimate variables from 51 species' occurrence records were tested. These 19 bioclimate variables were extracted from the corresponding layers using ArcGis 9.3. Only one variable from each set of highly cross-correlated variables

($r > 0.8$) was kept for further analysis. The decision of which variable was to be kept was based on both the correlation analysis and the PCA. For instance, the variables bio3 and bio2 were correlated ($r = 0.804$), and so were bio3 and bio4 ($r = -0.913$); considering the PCA result, bio3 was dropped, and both bio2 and bio4 were reserved. According to the contributions of 19 bioclimatic variables to every extracted principal component and the correlation analysis between the 19 bioclimatic variables, eight bioclimate variables (bio1, bio2, bio4, bio7, bio12, bio14, bio15, bio19) were extracted.

3) MAXENT Maximum Entropy (MaxEnt) Model

In 1957, Jaynes proposed maximum entropy theory, the essence of which is that on the basis of partial knowledge, the most reasonable inference about the unknown probability distribution is the most uncertain or the most random inference which matches the known knowledge. This is the only unbiased choice we can make; any other choice would introduce other constraints and assumptions that cannot be derived from the information we have.

The application of maximum entropy theory in species habitat suitability prediction can be expressed as follows: If we know nothing about a species' life habit or local ecological conditions, the most reasonable prediction is that the probabilities of the area being suitable for the species and of the area not being suitable are both 0.5. Any data that a species is present within a set of local ecological conditions is information that reduces the uncertainty of a MaxEnt model. The more information there is, the more uncertainty is reduced. The MaxEnt approach is to establish a model with a maximum entropy in accordance with known knowledge. Based on maximum entropy theory, the Java-based software package MaxEnt, which can be used for habitat suitability simulation, was developed by Phillips et al. The MaxEnt (version 3.3.3) we used in this study was obtained from http://www.cs.princeton.edu/~schapire/MaxEnt/ and can be downloaded freely for scientific research. The training data was 75% of the sample data selected randomly, and the test data was the remaining 25% of the sample data. The habitat suitability curves of each variable were calculated, and the contributions of each variable to the habitat model of H. riparia Lour were calculated using the software's built-in jackknife test.

There are four possible prediction results of the model: (i) the species exists where predicted to exist (true positive, TP); (ii) the species does not exist where predicted to exist (false positive, FP); (iii) the species exists where not predicted to exist (false negative, FN); (iv) the species does not exist where not predicted to exist (true negative, TN). Generally, both FN and FP errors always occur in the prediction results of SDM. These two types of errors both relate to the threshold that is used to determine presence or absence. Indexes frequently used for the evaluation of SDM performance are calculated based on true positive, false positive, true negative and false negative rates, including Cohen's Kappa, TSS (true skill statistic), AUC (area under ROC (receiver operating characteristic curve)) and others.

A number of different thresholds are set to calculate a series of sensitivities (positive rate in the positive results, TPTP+FP) and specificities (negative rate in the negative results, TNTN+FN). Plotting sensitivity on the ordinate against 1-specificity on the abscissa gives the ROC. The larger the AUC is, the better the model performance is. The AUC is not affected by choice of threshold, so it is an excellent index to evaluate model performance. The software package MaxEnt employs the AUC to evaluate model performance. In general, AUC is between 0.5 and 1. AUC $<$ 0.5 describes models that perform worse than chance and occurs rarely in reality. An AUC of 0.5 represents pure guessing. Model performance is categorized as failing (0.5—0.6), poor (0.6—0.7), fair (0.7—0.8), good (0.8—0.9), or excellent (0.9—1). The closer the AUC is to 1, the better the model performance is.

http://www.sciencedirect.com/science/article/pii/S0925857416302245

Vocabulary

H. riparia Lour　河岸滨藜
threshold 英 [ˈθreʃhəʊld] 美 [ˈθreʃhoʊld] n. 门槛；开始；[物] 临界值
specificity 英 [ˌspesɪˈfɪsəti] 美 [ˌspesɪˈfɪsəti] n. 特征；种别性；种特性

Useful Expressions

1. projection 英 [prəˈdʒekʃn] 美 [prəˈdʒekʃən] n. 预测；设计；突起物
They took me into a projection room to see a picture.
他们把我带到放映室去看一张图。
2. bioclimate 英 [biːəʊkˈlaɪmət] 美 [biːoʊkˈlaɪmət] n. 生物气候
The bioclimate analysis system was developed in Delphi 7.0, based on WinXP.
生物气候分析系统是在 WinXP 平台上，运用 Delphi 7.0 语言开发研制的。

Questions

1. Where is the data from?
2. How were variables selected?
3. What are the four possible prediction results of the models?

9.2 Main Contents of Medicinal Plant Resources Investigation

Passage 1

Biochemical Investigaton of the Plant Terminalia Arjuna (Part 1)

General Introduction

The study of disease and their treatment have been existing since the beginning of human civilization. Norman R. Farnsworth of the University of Illinois declared that, for every disease that affects mankind there is a treatment and cure occurring naturally on the earth. Plant kingdom is one of the major search areas for effective works of recent days. The importance of plants in search of new drugs is increasing with the advancements of medical sciences. For example, ricin, a toxin produced by the beans of Ricinus communis, has been found to be effectively couple to tumor targeted monoclonal antibiotics and has proved to be a very potent antitumor drug. Further have the HIV inhibitory activity has been observed in some novel coumarins (complex angular pyranocoumarins) isolated from Calophyllum lanigerum and glycerrhizin (from Glycerrhiza species). Hypericin (from Hypercium species) is an anticancer agent. Taxol is another example of one of the most potent antitumor agents found from Taxus bravifolia.

In fact, plants are the important sources of a diverse range of chemical compounds. Some of these compounds possessing a wide range of pharmacological activities are either impossible or to difficult to synthesize in the laboratory. A phytochemist uncovering these resources is producing useful materials for screening programs for drug discovery. Emergence of newer disease also leads the scientists to go back to nature for newer effective molecules.

Recently developed genetic engineering in plants has further increased their importance in the field of medicine, for example, in the production of antibiotics by expression of an appropriate gene in the plant. By using these techniques, it is possible to modify the activity or regulate the properties of the key enzymes responsible for the production of secondary metabolites. Thus by knowing the potential resources it is possible to increase the content of the important active compounds and in the future, genes responsible for very specific biosynthetic processes may be encoded into host organism to facilitate difficult synthetic transformation.

Thus plants are considered as are of the most important and interesting subjects that should be explored for the discovery and development of newer and safer drug candidates

1) Definition of Medicinal Plants

The plants that possess therapeutic properties or exert beneficial pharmacological effects on the animal body are generally designated as medicinal plants. Although there are no apparent morphological characteristics in the medicinal plants that make them distinct from other plants growing with them, yet they possess some special qualities or virtues that make them medicinally important. It has now been established that the plants, which naturally synthesize and accumulate some secondary metabolites like alkaloids, glycosides, tannins and volatile oils and contain minerals and vitamins, possess medicinal properties.

Accordingly the WHO consultative group on medicinal plants has formulated a definition of medicinal plants in the following way: " A medicinal plant is any plant which, in one or more of its organs, contains substances that can be used for therapeutic purposes or which are precursors for the synthesis of useful drugs. "

2) History of Medicinal Plants

From historical records, it is apparent that most of the early peoples like Assyrians, Babyloins Egyptians and ancient Hebrews, were familiar with the properties of many medicinal plants. Babyloins were aware of a large number of medicinal plants and their properties. Historical records of Assyria and Babyloins indicate that by 650 BC there were about 250 plant medicines in use in that region. As evident from the Papyrus Ebers, the ancient Egyptians possessed a good knowledge of medicinal properties of hundreds of plants. Many of the important plant drugs in present days like henbane (Hyoscymus spp.), mandrake (Mandragora officinarum). The great treatise supplies various information on medicinal use of the plants in the Indian subcontinent. It noted that the Indo-Aryans used the soma plants (Amaniat a muscaria and hallucinogenic mushroom) as medicinal agents. The practice of the medicine using medicinal plants flourished most during the Greek civilization when historical personalities like Hippocrates and Theophrastus practice herbal medicine. The materia medica of the Greek physician Hippocrates consists of some 300—400 medicinal plants which include opium, mint, rosemary, sage and verbena. The current list of the medicinal plants growing around the world includes more than a thousands items.

3) Contribution of Medicinal Plants to Modern Medicine

Medicinal plants have been serving as the major sources of medicine for maintenance of health and wellbeing of the human beings from the very beginning of their existence on earth. These medicinal plants were used by the early man, as are done now, in a variety of forms, such as in the entire form, and as powders, pastes, juices, infusions and decoctions for the treatment of their various disease and ailments. These various converted forms of the medicinal plants may thus be very conveniently and genuinely called medicinal preparations or medicaments.

In this way, the medicinal plants formed an integral part of the health management practices and constituted important items of medicines from the very early days of human civilization. In the course of their traditional uses, the medicinal plants have contributed substantially to the gradual development of medicines to their present state. As therapeutic uses of plants continued with the progress of civilization and development of human knowledge, scientists endeavored to isolate different chemical constituents from plants, put them to biological and pharmacological tests and thus have been able to get prepared. In this way, ancient uses of Datura plants have led to the isolation of hyoscine, hyoscyamine, atropine and tigloidine, Cinchona baek to quinine and quinidine, Rauvilfia serpentineto reserpine and rescinnamine, Digitalis purpurea to digitoxin and digoxin, papaver somniferum to morphine and codaine, Ergotamine and ergometrine, Senna to sennosides. Catharanthus roseus to vinblastine and vincristine—to mention a few. Isolation of the natural analgesic drug to morphin from papaver somniferumcapsules, in 1804 is probably the first most important example of naturadrug to which plants have directly contributed.

http://www.lawyersnjurists.com/article/biochemical-investigation-plant-terminalia-arjuna-arjun/

Vocabulary

ricin 英 ['raɪsɪn] 美 ['raɪsɪn, 'rɪsɪn] *n.* 篦麻毒素；篦麻蛋白
antibiotics 英 [ˌæntɪbaɪ'ɒtɪks] 美 [ˌæntɪbaɪ'ɒtɪks] *n.* （用作复数）抗生素；（用作单数）抗生物质的研究
Taxus bravifolia 短叶红豆杉
biosynthetic 英 [biːəʊ'sɪnθetɪk] 美 [biːoʊ'sɪnθetɪk] *adj.* [医] 生物合成的
therapeutic property 治疗性能
henbane 英 ['henbeɪn] 美 ['hen,beɪn] *n.* 茄科的药用植物；天仙子
mandrake 英 ['mændreɪk] 美 ['mæn,dreɪk] *n.* 曼德拉草；曼德拉草根（旧时用作药物）
soma plant 胞体植物
verbena 英 [vɜː'biːnə] 美 [vɜːr'biːnə] *n.* 马鞭草属植物
analgesic drug morphin 镇痛药物吗啡

Useful Expressions

1. monoclonal 英 [ˌmɒnəʊ'kləʊnəl] 美 [ˌmɒnoʊ'kloʊnəl] *adj.* 单克隆的
It may be beneficial to the future research on properties and applications of the monoclonal antibodies.
这为单抗的性质和应用的深入研究打下了基础。
2. treatise 英 ['triːtɪs] 美 ['trɪtɪs] *n.* 论述；论文；专著
The Classics on Tea, written by Lu Yu of the Tang Dynasty(618 AD—907 AD) was the world's earliest treatise on tea leave production.

中国唐代陆羽写的《茶经》是世界上最早的制茶专著。

3. sage 英 [seɪdʒ] 美 [sedʒ] n. 圣人；贤人；智者；鼠尾草（可用作调料）

He was famous for his sage advice to younger painters.

他因向后起的年轻画家提供睿智的忠告而闻名。

Questions

1. What does Norman R. Farnsworth of the University of Illinois say about disease?
2. What are plants regarded as? Why?
3. What can be called medicinal plants?
4. What do medicinal plants contribute to modern medicine?

Passage 2

Biochemical Investigaton of the Plant Terminalia Arjuna (Part 2)

Plants Preiew

1) Terminalia Arjuna

Terminalia arjuna is a medicinal plant of the genus Terminalia, widely used by ayurvedic physicians for its curative properties in organic/functional heart problems including angina, hypertension and deposits in arteries. According to Ayurvedic texts it is also very useful in the treatment of any sort of pain due a fall, ecchymosis, spermatorrhoea and sexually transmitted diseases such as gonorrhoea. Arjuna bark (Terminallia arjuna) is thought to be beneficial for the heart. This has also been proved in a research by Dr. K. N. Udupa in Banaras Hindu University's Institute of Medical Sciences, Varanasi (India). In this research, they found that powdered extract of the above drug provided very good results to the people suffering from Coronary heart diseases Research suggests that Terminalia is useful in alleviating the pain of angina pectoris and in treating heart failure and coronary artery disease. Terminalia may also be useful in treating hypercholesterolemia. The cardio protective effects of terminalia are thought to be caused by the antioxidant nature of several of the constituent flavonoids and oligomeric proanthocyanidins, while positive inotropic effects may be caused by the saponin glycosides. In addition to its cardiac effects, Terminalia may also be protective against gastric ulcers, such as those caused by NSAIDs.

The leaves of this tree are also fed on by the Antheraea paphia moth which produces the tassar silk (Tussah), a form of wild silk of commercial importance.

(1) Description

Terminalia arjuna is a deciduous tree found throughout India growing to a height of 60—90 feet. The thick, white-to-pinkish-gray bark has been used in India's native Ayurvedic medicine for over three centuries, primarily as a cardiac tonic. Clinical evaluation of this botanical medicine indicates that it can be of benefit in the treatment of coronary artery disease, heart failure, and possibly hypercholesterolemia. It has also been found to be antiviral and antimutagenic.

(2) Botany

Tree up to 25-m high; bark grey, smooth; leaves sub-opposite, oblong or elliptic oblong, glabrous, often inequilateral, margin often crenulate, apex obtuse or sub-acute, base rounded or sometimes cordate; petioles 0.5—1.2 cm. Flowers are small and white. Fruits are 2.3—3.5-cm long, fibrous woody, glabrous with 5 hard wings, striated with numerous curved veins. Flowering time is from April to July in Indian conditions. Seeds can be kept under hard germination for 50—76 days (50%—60%).

(3) Useful Parts

Every part with useful medicinal properties of Arjun holds a reputed position in both Ayurvedic and Yunani systems of medicine. According to Ayurveda it is alexiteric, styptic, tonic, anthelmintic, and useful in fractures, uclers, heart diseases, biliousness, urinary discharges, asthma, tumours, leucoderma, anaemia, excessive prespiration, etc. According to Yunani system of medicine, it is used both externally and internally in gleet and urinary discharges. It is used as expectorant, aphrodisiac, tonic and diuretic.

(4) Traditional Uses

Every part of the tree has useful medicinal properties. Arjun holds a reputed position in both Ayurvedic and Yunani Systems of medicine. According to Ayurveda it is alexiteric, styptic, tonic, anthelmintic, and useful in fractures, uclers, heart diseases, biliousness, urinary discharges, asthma, tumours, leucoderma, anaemia, excessive prespiration, etc. According to Yunani system of medicine, it is used both externally and internally in gleet and urinary discharges.

(5) Therapeutic Properties and Uses

The bark constituents an important crude drug, which is esteemed as a cardiac tonic. It reduces blood pressure and cholesterol levels. Powdered bark relives hypertension. It has a diuretic and general tonic effect in case of liver cirrhosis. The bark also acts as an astringent and febrifuge, and is used in the treatment of red and swollen mouth, tongue and gums. It stops bleeding and pus formation in the gums, and is useful in asthma, dysentery, menstrual problems, pain, leucorrhoea, wounds and skin eruptions. Cardiomyopathy like Myocardial infraction, angina, coronary artery disease, heart failure, hypercholesterolemia, hypertension. In case of heart attack though it can not act against, like streptokinase or eurokinase, but regular use of it after just recovering from heart attack, reduces the chance of further attack to a great level. Besides, no such toxicity or side effects has/have been found so far, which can be advocated to use in regular basis for a strong and well functioning heart. Arjuna reduces angina episodes much better than nitroglycerin. In one study, angina episodes were cut in half by the Arjuna, with none of the nasty side effects. Plus, it can be used as long as you like, and it'll stop working. Arjuna has been shown to help reverse hardening of the arteries.

(6) Chemical Constituents

Arjuna bark contains tannins (12%), sitosterol, triterpenoid saponins (arjunic acid, arjunolic acid, arjungenin, arjunglycosides), essential oil, reducing sugars, calcium salts and traces of aliminium and magnisun salts, flavonoids (arjunone, arjunolone, luteolin), gallic acid, ellagic acid, oligomeric proanthocyanidins (OPCs), phytosterols, calcium, magnesium, zinc, and copper.

(7) Mechanisms of Action

Improvement of cardiac muscle function and subsequent improved pumping activity of the heart seems to be the primary benefit of Terminalia. It is thought the saponin glycosides might be responsible for the inotropic effect of Terminalia, while the flavonoids and OPCs provide free radical antioxidant activity and vascular strengthening. A dose-dependent decrease in heart rate and blood pressure was noted in dogs intravenously. Recently, two new cardenolide cardiac glycosides have been isolated from the root and seed of Terminalia. The main action of these cardenolides is to increase the force of cardiac contraction by means of a rise in both intracellular sodium and calcium.

2) Clinical Indications

(1) Angina Pectoris

An open study of Terminalia use in stable and unstable angina demonstrated a 50-percent reduction of angina in the stable angina group after three months ($p<0.01$). A significant reduction was also found in systolic blood pressure in these patients ($p<0.05$). During treadmill testing, both the onset of angina and the appearance of ST-T changes on ECG were significantly delayed in the stable angina group ($p<0.001$), indicating an improvement in exercise tolerance. The unstable angina group did not experience significant reductions in angina or systolic blood pressure. Both groups showed improvements in left ventricular ejection fraction. Evaluation of overall clinical condition, treadmill results, and ejection fraction showed improvement in 66% of stable angina patients and 20 percent of unstable angina patients after three months. In this study Terminalia was also associated with a lowering of systolic blood pressure. Two clinical studies found similar results when Terminalia arjuna was compared to isosorbide mononitrate in stable angina patients. Both studies showed a similar reduction in the number of anginal episodes, as well as improvements in stress tests. In one study 58 males with chronic stable angina with evidence of ischemia on treadmill testing received Terminalia.

Arjuna (500 mg every eight hours), isosorbide mononitrate (40 mg daily), or a matching placebo for one week each, separated by a wash-out period of at least three days in a randomized, double-blind, crossover design. Terminalia therapy was associated with a significant decrease in the frequency of angina and need for isosorbide dinitrate. Treadmill parameters improved significantly during therapy with Terminalia compared to those with placebo. Similar improvement in clinical and treadmill parameters were observed with isosorbide mononitrate compared to placebo therapy. No significant difference was observed in clinical or treadmill parameters when Terminalia arjuna and isosorbide mononitrate therapies were compared.

(2) Congestive Heart Failure

A double-blind, placebo-controlled, two-phase trial of Terminalia extract in 12 patients with severe refractory heart failure (NYHA Class IV) was conducted, in which either 500 mg Terminalia bark extract or placebo was given every eight hours for two weeks, in addition to the patients' current pharmaceutical medications (digoxin, diuretics, angiotensin-convertingenzyme inhibitors, vasodilators, and potassium supplementation). All patients experienced dyspnea at rest or after minimal activity at the start of the trial. Dyspnea, fatigue, edema, and walking tolerance were all improved while patients were on Terminalia therapy. Treatment with Terminalia was also associated with significant improvements in stroke volume and left ventricular ejection fraction, as well as decreases in end-diastolic and end-systolic left ventricular volumes compared to placebo. In the second phase of the study, patients from phase I continued on Terminalia extract for two years. Improvements were noted in the ensuing 2—3 months, and were maintained through the balance of the study. After four months' treatment, nine patients improved to NYHA Class II and three improved to NYHA Class III.

(3) Cardiomyopathy/Post-Myocardial Infarction

A study was conducted on 10 post-myocardial-infarction patients and two ischemic cardiomyopathy patients, utilizing 500 mg Terminalia extract every eight hours for three months, along with conventional treatment. Significant reductions in angina and left ventricular mass, in addition to improved left ventricular ejection fraction, were noted in the Terminalia group; whereas, the control group taking only conventional drugs experienced decreased angina only. The two patients with cardiomyopathy were improved from NYHA Class III to NYHA Class I during the study.

(4) Hyperlipidemia

Animal studies suggest Terminalia might reduce blood lipids. Rabbits made hyperlipidemic on an atherogenic diet were given an oral Terminalia extract, and had a significant, dose-related decrease in total- and LDL-cholesterol, compared to placebo ($p < 0.01$) However, the amounts used (100 mg/kg and 500 mg/kg body weight) were very large, and it remains to be seen if similar changes will be observed in humans taking relatively smaller oral doses. In a similar study of rats fed cholesterol (25 mg/kg body weight) alone or along with Terminalia bark powder (100 mg/kg) for 30 days, Terminalia feeding caused a smaller increase in blood lipids and an increase in HDL cholesterol, compared to the cholesterol-only group. The researchers concluded that inhibition of hepatic cholesterol biosynthesis, increased fecal bile acid excretion, and stimulation of receptor-mediated catabolism of LDL cholesterol were responsible for Terminalia's lipid-lowering effects.

In another study, rabbits were fed a cholesterol-rich diet in combination with three indigenous terminalia species; Terminalia arjuna, Terminalia belerica, and Terminalia chebula. Upon histological examination, the rabbits fed the diet and Terminalia arjuna exhibited the most potent hypolipidemic effect, with partial inhibition of atheroma.

In a randomized, controlled trial, Terminalia bark was compared to vitamin E. 105 patients with coronary heart disease (CHD) were matched for age, lifestyle, and diet variables, as well as drug treatment status. None of the patients were previously on lipid-lowering medications. Placebo, vitamin E (400 IU), and Terminalia (500 mg) were administered.

Results showed no significant changes in the placebo group or the vitamin E group. The Terminalia group had a significant decrease in total cholesterol and LDL cholesterol. Lipid peroxidase levels decreased significantly in both vitamin E and Terminalia groups; however, there was a greater decrease in the vitamin E group.

3) Other Clinical Indications

Terminalia bark harbors constituents with promising antimutagenic and anticarcinogenic potential that should be investigated further. *in vitro* studies have also shown that Terminalia possesses anti-herpes virus activity.

4) Botanical-Drug Interactions

Terminalia arjuna extracts have been used in clinical studies concomitantly with standard heart medications, including digoxin, diuretics, angiotensin-converting-enzyme inhibitors, and vasodilators, with no reported adverse effects. Simultaneous use of Terminalia with other cardiac medications should be undertaken with caution.

5) Dosage and Toxicity

A typical dose of dried bark is 1—3 grams daily, while 500 mg bark extract four times per day has been used in congestive heart failure. No toxicity has been documented.

6) Pharmacology

The bark is acrid, and credited with styptic, tonic, febrifugal and antidysenteric properties. In fractures and contusions, with excessive ecchymosis, the powdered bark is taken with milk. The powdered bark seemed to give relief in symptomatic complaint in hypertension; it apparently had a diuretic and a general tonic effect in cases of cirrhosis of the liver. A decoction of the bark is used as a wash in ulcers. The alcoholic extract of bark contained CaO, 0.33; MgO, 0.078; and Al_2O_3, 0.076. The fruit was tonic and deobstruent. The juice of the fresh leaves was used in earache. These extracts also inhibited carotid occlusion response. Hypotension and bradycardia were also observed and were mainly of central origin. Arjuna bark was popularly used as a cardiac tonic. The diuretic property was also observed, because of a saponin "The bark is useful in diseases of the heart, allays thirst and relieves fatigue".

7) Present Study Protocol

Our present study was designed to isolate pure compounds as well as to observe pharmacological activities of the isolated pure compounds with crude extracts of the plant Terminalia arjuna. The study protocol consisted of the following steps:

(i) Extraction of the powdered bark of the plant with distilled methanol;

(ii) Filtration of the crude methanolic extract by using the marking cotton cloth and subsequently through the filter paper and solvent evaporation;

(iii) Elucidation of the structure of the isolated compounds with the help of spectroscopic method;

(iv) Brine shrimp lethality bioassay and determination of LC_{50} of crude extract;

(v) Investigation of *in vitro* antimicrobial activity of crude extract;

(vi) Determination of antioxidant activity of crude extract.

http://www.lawyersnjurists.com/article/biochemical-investigation-plant-terminalia-arjuna-arjun/

Vocabulary

Terminalia arjuna　阿江榄仁树

Terminalia　*n.* 榄仁树属

ayurvedic physician　印度草医药学

angina 英 [æn'dʒaɪnə] 美 [æn'dʒaɪnə, 'ændʒə-] *n.* 心绞痛；咽峡炎；咽喉痛

hypertension 英 [ˌhaɪpə'tenʃn] 美 [ˌhaɪpər'tenʃn] *n.* 高血压；过度紧张

deposits in arteries　动脉沉积物

ecchymosis 英 [ˌekɪ'məʊsɪs] 美 [ˌekə'moʊsɪs] *n.* 出血斑；瘀癍

spermatorrhoea 英 [ˌspɜːmətə'riːə] 美 [ˌspɜːmətə'riːə] *n.* 遗精

gonorrhea 英 [ˌɡɒnə'riːə] 美 [ˌɡɒnə'riːə] *n.* 淋病；白浊

angina pectoris 英 [æn'dʒaɪnə'pektərɪs] 美 ['pektərɪs] *n.* 心绞痛；真心痛

coronary artery disease　冠心病

hypercholesterolemia 英 [ˌhaɪpə(ː)kəlestərəʊ'liːmjə] 美 [ˌhaɪpəkəˌlestərə'liːmiːə] *n.* 血胆脂醇过多；血胆甾醇过多；血胆固醇过多症

cardio protective effects of terminalia　诃子的心脏保护作用

cardiac tonic　[化]治疗心功能不全药；强心药；[医]强心剂

styptic 英 ['stɪptɪk] 美 ['stɪptɪk] *n.* 止血剂

anthelmintic 英 [ˌænt'helmɪntɪk] 美 [ˌənt'helmɪntɪk] *n.* 驱虫剂；打虫药

liver cirrhosis　慢性间质性肝炎

febrifuge 英 [ˈfebrɪfjuːdʒ] 美 [ˈfebrɪˌfjuːdʒ] n. 解热药；退热药
pus formation   脓液形成
dysentery 英 [ˈdɪsəntri] 美 [ˈdɪsənteri] n. 痢疾；脏毒
skin eruption   皮疹
artery 英 [ˈɑːtəri] 美 [ˈɑːrtəri] n. [解剖] 动脉；干线；要道
refractory 英 [rɪˈfræktəri] 美 [rɪˈfræktəri] adj. 难治疗的；耐熔的
antimutagenic 英[ˌæntɪmjuːtɪdˈʒenɪk] 美 [ˌæntɪmjuːtɪdˈʒenɪk] adj. 抗诱变剂的；抗诱变因素的
congestive heart failure   充血性心力衰竭
pharmacology 英 [ˌfɑːməˈkɒlədʒi] 美 [ˌfɑːrməˈkɑːlədʒi] n. 药理学；药物学

Useful Expressions

1. relive 英 [ˌriːˈlɪv] 美 [riˈlɪv] vt.（在想象中）重新过……的生活；再经历
There is no point in reliving the past.
回味过去没有任何意义。
2. ventricular 英 [venˈtrɪkjʊlə] 美 [venˈtrɪkjələ] adj. 心室的；膨胀的
TIMP-1 may promote ventricular remodeling and accelerate the process of the myocardial diastolic dysfunction in hypertensive heart disease.
TIMP-1 可能促进了高血压患者心室重构，加快了心肌舒张功能障碍的进程。

Questions

1. Why is Terminalia arjuna used widely?
2. Where can Terminalia be found?
3. How tall is Terminalia?
4. What can the bark of Terminalia cure?

9.3 Application of Modern Information Technology in Investigation of Medicinal Plant Resources

Passage 1

Systems Medicine: The Application of Systems Biology Approaches for Modern Medical Research and Drug Development (Extract)

Introduction

The exponential development of highly advanced scientific and medical research analytical technologies throughout the past 30 years has arrived to the point where most (if not all) key molecular determinants deemed to affect human conditions and diseases can be scrutinized with great detail.

Scientists and clinicians can now begin to attempt investigation of any individual dysregulation occurring within the genomic, transcriptomic, miRnomic, proteomic, and metabolomic levels thanks to advancing wet-lab technologies such as mass spectrometry, quantitative polymerase chain reaction (QPCR) and next generation sequencing, and detailed bioinformatics suites. All these technologies are capable of extracting information from complex datasets to enable disease models to be developed for wet-lab testing. The interplay between the wet and dry lab with specific clinical expertise not only is a main current component of translational medicine, but also is enabled by systems medicine.

However, there are drawbacks within this scientific brave new age, in that in most scientific studies it is only specific molecular levels which are individually investigated for their influence in affecting any particular health condition. Ideally, any form of medical research with the scope of rooting out dysregulated molecular pathway interactions should focus on investigating the holistic aspects of the complex and multifactorial medical conditions. This involves careful and methodical examination of all simultaneous molecular interactions occurring levels (e. g. genomic, transcriptomic, etc.). Such "bigger-picture" research perspectives lead to a higher level of understanding for complex and multifactorial disease conditions and ultimately "fast-track" the identification and clinical diagnosis of specific molecular pathway dysregulations with pathogenesis value, together with the combined identification of novel drug targets for the development of effective translational therapeutics for the medical condition.

Consequently, the urgent need to counteract such research shortcomings has been acquiesced through the emergence, in the last decade, of the novel research field of systems biology.

Main Principles of Systems Biology Approaches to Research

In essence, the field of systems biology revolves around the principle that the phenotype of any individual living organism is a reflection of the simultaneous multitude of molecular interactions from various levels occurring at any one time, combined in a holistic manner to produce such a phenotype. Consequently, against the standard concept of reductionist approach where dysregulations in isolated molecular components are studied, data from dysregulations of multiple key molecular players from varying cellular levels of activity are pooled and studied in their entirety, for the purpose of identifying distinct changes in the pattern of intermolecular relationships, *vis a vis* the organism's investigated phenotypes. The methods applied as the principal research tools vary, depending on the nature of the molecular level being investigated and also on the volumes of data generated; therefore, nowadays most self-sufficient systems biology research groups are composed of research scientists who are unique experts in their own specific research field with a discernable knowledge of experimental investigation for most molecular level research.

Consequently, systems biology is an interdisciplinary field of research, requiring the technology platforms and research expertise of individuals from a spectrum of scientific research niche. However, the measurement of all molecular parts of an organ or even biomedical pathway is far from routinely achievable, and great efforts to improve sensitivity of analysis and to make the output data possess a quantitative significance have been made to improve through implementation of field standards. Given current constraints, Boolean approaches are assisting with production of first generation systems biology models. A main difference between systems biology and systems medicine is that the former assumes the data to be correct and usable as often wet-lab data generation expertise is not the main goal but is assumed to be correct and usable. Systems medicine (sometimes referred to as systems healthcare) promises to lead with clinical and molecular know-how to produce exquisite datasets that are employed to generate pathway models and treatment and will hopefully contribute to stratified medicine en route to personalized healthcare in a direct way.

In addition to performing function as an interdisciplinary research field, systems biology research methods rely heavily on the bioinformatics/computational and mathematical modeling components for achieving answers to the specific research questions. Such informatics technology utilization can be twofold in systems biology, namely, the implementation of a hypothesis driven "top-down" approach or experimental data driven "bottom-up" approaches.

The bottom-up, data driven approach initiates from the collection of large volume datasets derived through a spectrum of omics-based experimental procedures, followed by thorough mathematical modeling analyses to combine the relationships between key molecular players from the varying omics data results obtained. One of the primary methodologies employed by the bottom-up systems biology conceptual approach is network modeling. A typical biological network model is composed of multiple nodes interacting with each other through edges, whereby nodes are classified as individual key molecular players from any omics level, such as genes, noncoding RNA family members, and proteins and the edges represent experimentally validated molecular interactions. Both the nature and detail of the nodes and edges within any particular biological network may vary. In addition, highly active nodes interacting in a close-knit network are defined as hubs. Hubs can be further subdivided into two categories, namely, "party" hubs and "date" hubs. Party hubs represent nodes which commonly interact with multiple other molecular partners in a simultaneous manner, whereas date hubs are much more dynamic since they interact with other molecular partners across multiple timeframes and within varying locations.

Conversely to the bottom-up experimental methodologies, the hypothesis driven top-down approach relies heavily on mathematical modeling for conducting studies on small-scale molecular interactions for a specific biological condition or phenotype. The dynamical modeling employed for such purpose involves the translation of molecular pathway interactions present in the studied organism into defined mathematical formats, such as ordinary differential equations (ODEs) and partial differential equations (PDEs) that can be analysed and probed within a "dry lab" environment. Such a method can be utilized since most intermolecular activities occur with specific kinetics that can be mimicked (e. g. Michaelis-Menten kinetics) by appropriate mathematical derivations. However, dynamical modeling can only be effective if specific assumptions are imposed regarding the biomolecular interactions taking place, such as the selection of defined reaction rate kinetics occurring within the studied biomolecular interactions.

In summary, there are four main phases to develop accurately functioning dynamical modeling, namely, model design to identify the pillar intermolecular activities, model construction of such molecular interactions into representative differential equations, model calibration to identify and modulate nonspecific kinetics of individual biomolecular components of the model for the purpose of fine tuning the mathematical model to the experimental format, and model validation by inferring distinct predictions that can be verified in a "wet-lab" experimental scenario.

Interestingly, there can also be a third approach to systems biology research models that implement both the top-down and bottom-up methodologies, namely, the middle-out (rational) approach.

https://www.hindawi.com/journals/mbi/2015/698169/

Vocabulary

exponential 英 [ˌekspəˈnenʃl] 美 [ˌɛkspəˈnɛnʃəl] *adj.* 越来越快的
clinician 英 [klɪˈnɪʃn] 美 [klɪˈnɪʃən] *n.* 临床医生；门诊医师
dysregulation 英 [diːzregjʊˈleɪʃn] 美 [diːzregjʊˈleɪʃn] *n.* [医] 失调
spectrometry 英 [spekˈtrɒmɪtrɪ] 美 [spekˈtrɒmɪtrɪ] *n.* 光谱测定法；谱测量
bioinformatics suite 生物信息套件
interplay 英 [ˈɪntəpleɪ] 美 [ˈɪntərpleɪ] *n.* 相互作用
dysregulate 英 [ˌdɪsˈregjuleɪt] 美 [dɪsˈrɛgjəˌlet] *vt.* 增生调节；异常调节
methodical 英 [məˈθɒdɪkl] 美 [məˈθɑːdɪkl] *adj.* 有条不紊的；有方法的
reductionist 英 [rɪˈdʌkʃənɪst] 美 [rɪˈdʌkʃənɪst] *n.* 还原论者
entirety 英 [ɪnˈtaɪərəti] 美 [ɛnˈtaɪrɪti, -ˈtaɪrti] *n.* 完全；整体
vis a vis [医] 对面；相对
interdisciplinary 英 [ˌɪntəˈdɪsəplɪnəri] 美 [ˌɪntərˈdɪsəplɪneri] *adj.* 跨学科；各学科间的
hypothesis 英 [haɪˈpɒθəsɪs] 美 [haɪˈpɑːθəsɪs] *n.* 假设；假说；[逻] 前提
intermolecular 英 [ˌɪntə(ː)məˈlekjʊlə] 美 [ˈɪntəməˈlekjələ] *adj.* 分子间的；存在（或作用）于分子间的

kinetics 英 [kɪˈnetɪks] 美 [kɪˈnetɪks] n. 动力学

Useful Expressions

1. determinant 英 [dɪˈtɜːmɪnənt] 美 [dɪˈtɜːrmɪnənt] n. 行列式；决定因素；决定物；免疫因子

The windows and the views beyond them are major determinants of a room's character.

窗户和窗外的景色是决定一个房间特色的主要因素。

2. deem 英 [diːm] 美 [diːm] vt. 认为；视为；主张；断定

French and German were deemed essential.

法语和德语被认为是必不可少的。

3. scrutinize 英 [ˈskruːtənaɪz] 美 [ˈskruːtnˌaɪz] vt. 仔细检查

Her purpose was to scrutinize his features to see if he was an honest man.

她的目的是通过仔细观察他的相貌以判断他是否诚实。

4. acquiesce 英 [ˌækwiˈes] 美 [ˌækwiˈɛs] vi. 默认；默许

Steve seemed to acquiesce in the decision.

史蒂夫好像默认了这个决定。

5. probe 英 [prəʊb] 美 [proʊb] vt. 调查；用探针（或探测器等）探查；探测

The more they probed into his background, the more inflamed their suspicions would become.

他们越调查他的背景，疑团就越多。

Questions

1. What involves careful and methodical examination of all simultaneous molecular interactions occurring levels?

2. What are main principles of systems biology approaches to research?

3. What are the four main phases to develop accurately functioning dynamical modeling?

Passage 2

A System-Level Investigation into the Mechanisms of Traditional Chinese Medicine: Compound Danshen Formula for Cardiovascular Disease Treatment (Extract)

Introduction

Cardiovascular diseases (CVDs) are the leading cause of death in the world. In 2008, about 17.3 million people died from CVDs, representing 30% of total global deaths. The number has been estimated to increase to 23.6 million by 2030. Although diverse drugs and medications have already been employed on CVDs, developing new therapeutic tools are still in urgent need and under intensive investigation. As one of these efforts, modernization of traditional Chinese medicine (TCM) has attracted a lot of attention.

Compound Danshen Formula (CDF) is one of TCM recipes for treatment of CVDs which is composed of Radix Salviae Miltiorrhizae (Labiatae sp. plant, Chinese name Danshen), Panax Notoginseng (Araliaceae plant, Chinese name Sanqi), and Borneolum (Crystallization of the resin and volatile oil in Cinnamomum camphora (L.) Presl, Chinese name Bingpian), at a ratio of 450:141:8. CDF is officially registered in Chinese Pharmacopoeia and has been widely used to treat CVDs in China, Japan, the United States and Europe. Clinical studies have revealed a variety of desirable pharmacological effects of CDF on CVDs, such as increasing coronary flow rate, activating superoxide dismutase, dilating coronary vessels, etc., which contributes significantly to the survival rate of CVDs patients. However, the molecular details about how CDF can be administrated on CVDs are still unclear.

Studies on CDF's pharmacological effect have confronted several major challenges. Firstly, isolation and identify chemical constituents possessing desirable pharmacological effects are labor-intensive, time-consuming and costly, given the fact that most medicinal herbs may contain tens of thousands constituents. Secondly, a certain ingredient may function on several relevant or irrelevant biological targets, which makes its pharmacological and toxicological effects difficult to be evaluated independently. Thirdly, and most importantly, TCM, such as CDF, has traditionally been administered as an integrated prescription for treating diseases which implicate that a complex, and highly dynamic ingredient-ingredient interaction network may underlying the overall clinical effect.

Systems pharmacology has emerged as a promising subject to overcome these challenges by providing powerful new tools and conceptions. Network analysis is one of these approaches which can evaluate TCM's pharmacological effect as a whole unity. In this work, we proposed for the first time a systems-pharmacological model by combining oral bioavailability prediction, multiple drug-target prediction and validation, and network pharmacology techniques, to shed new lights on the effectiveness and mechanism of CDF. Different types of data, such as the physiological, biochemical and genomic information have been collected to build the model which is based on an array of computational approaches including the machine learning method and network analysis. The proposed network-driven, integrated approach would also provide a novel and efficient way to deeply explore the chemical and pharmacological basis of TCM.

Materials and Methods

As a combination of three plants, CDF contains a considerable number of chemical compounds and some of which have been demonstrated to possess significant pharmacological activities. This provides an important basis to bring systems biology insights into the investigation of TCM theory and practice. In the following part, we will introduce how to build database and models for this CDF.

Database Construction

In this study all chemicals of each herb were retrieved from Chinese Academy of Sciences Chemistry Database and Chinese Herbal Drug Database and literature. Finally, to the most extent 320 compounds were collected, including 201 in Radix Salviae Miltiorrhizae, 112 in Panax Notoginseng and 31 in Borneolum, respectively (the three herbal shared the same 24 compounds). The structures of these molecules were downloaded from LookChem or produced by ISIS Draw 2.5 and further optimized by Sybyl 6.9 with sybyl force field and default parameters. The molecules were saved to a mol format for further analysis.

In our previous work, we developed a robust in silico model OBioavail 1.1, which integrated with the metabolism information to predict a compound human oral bioavailability. The model was built based on a set of 805 structurally diverse drugs and drug-like molecules which have been critically evaluated for their human oral bioavailability. The multiple linear regression, partial least square and support vector machine (SVM) methods were employed to build the models, resulting in an optimal model with $R2=0.80$, $SEE=0.31$ for the training set, $Q2=0.72$, $SEP=0.22$ for the independent test set. In this work, the compounds with OB $\geqslant 50\%$ were selected as the candidate compounds. The threshold determination is based upon the careful consideration of the following rules: (i) extracting information as much as possible from CDF using the least number of compound; (ii) the obtained model can be reasonably interpreted by the reported pharmacological data.

Target Identification

The targets were searched by PharmMapper Server, which is designed to identify potential target candidates for the given small molecules (drugs, natural products, or other newly discovered compounds with targets unidentified) via a 'reverse' pharmacophore mapping approach. The model is supported by a large repertoire of pharmacophore database composed of more than 7,000 receptor-based pharmacophore models that are extracted from TargetBank, DrugBank, BindingDB and PDTD. A strategy algorithm of sequential combination of triangle hashing and genetic algorithm optimization is designed to solve the molecule pharmacophore best fitting task. In this work, the number of the reserved matched targets is defined as 300 with the fitting score $\geqslant 3.00$. The target set is only limited to the human targets (2,214); and all parameters are kept as default. The information of the predicted target candidates which have relationships with CVD is collected and further verified from TTD, PharmGkb and DrugBank.

Target Validation

Docking: To validate the compound-target associations related with CVDs, the molecular docking simulation was further performed on each bioactive compound complexed with their human target enzymes by AutoDock software (version 4.2). All the protein structures except P-glycoprotein (P-gp) were directly downloaded from the RCSB protein data bank (www.pdb.org) with their resolutions being carefully checked. The homology model of P-gp was obtained from our previous work. AutoDock tools (ADT) (version 1.4.5) were used for protein and ligand preparation. Generally, all hydrogens, including non-polar, Kollman charges and solvation parameters were added to individual molecules. For all ligands, the Gasteiger charges were assigned with the nonpolar hydrogens merged. The auxiliary program Autogrid was used to generate the grid maps for each sample. The docking area was defined by a $60 \times 60 \times 60$ 3D grid centered around the ligands binding site with a 0.375 Å grid space. All bond rotations for the ligands were ignored and the Lamarckian genetic algorithm (LGA) was employed for each simulation process.

Molecular Dynamics Simulation

All molecular dynamics simulations were carried out using the Amber 10 suite of programs. The standard AMBER99SB force field was selected for proteins, the ligand charges and parameters were determined with the antechamber module of Amber based on the AM1-BCC charge scheme and the general atom force field (GAFF). All models were solvated in the rectangular box of TIP3P water extending at least 10 Å in each direction from the solute, and neutralized by adding sufficient Na^+/Cl^- counterions. The cut-off distance was kept to 8 Å to compute the nonbonded interactions. All simulations were performed under periodic boundary conditions, and the long-range electrostatics were treated by using the Particle-mesh-Ewald method (PME). All bonds containing hydrogen atoms were fixed using the SHAKE algorithm.

After initial configuration construction, a standard equilibration protocol was performed for MD simulations. The systems were minimized by 500 steps of steepest descent and 1,000 steps of conjugate gradient to remove the bad contacts in the structure, then were slowly heated to 300 K over 50 ps using 2.0 kcal/mol/Å-2 harmonic restraints. Subsequently, a 50 ps pressure-constant (1 bar) period to raise the density while still keeping the complex atoms constrained and a 500 ps equilibration were conducted. The production stage consisted of a total of 5 ns at constant temperature of 300 K for each system, respectively. The integration time step was 2 fs and the coordinates were saved every 2 ps.

http://journals.plos.org/plosone/article?id=10.1371/*journal.pone*.0043918

Vocabulary

medication 英 [ˌmedɪˈkeɪʃn] 美 [ˌmɛdɪˈkeʃən] *n.* 药物治疗；药物
Compound Danshen Formula 复方丹参方
superoxide 英 [sjuːpəˈɒksaɪd] 美 [sjuːpəˈɒksaɪd] *n.* 过氧化物；超氧化物
coronary vessel 冠状血管
bioavailability 英 [ˌbaɪəʊəˌveɪləˈbɪlɪtɪ] 美 [ˌbaɪoʊrˌveɪləˈbɪlɪtɪ] *n.* （药物或营养素的）生物药效率；生物利用度
vector 英 [ˈvektə(r)] 美 [ˈvɛktɚ] *n.* 矢量；航向；[生] 带菌者
pharmacophore 英 [ˈfɑːməkəfɔː] 美 [ˈfɑːməˈkoufɔː] *n.* 药效团
ligand 英 [ˈlɪɡənd] 美 [ˈlɪɡənd] *n.* 配体；配合基；向心配合（价）体
rectangular 英 [rekˈtæŋɡjələ(r)] 美 [rɛkˈtæŋɡjələ˞] *adj.* 长方；[数] 矩形的；成直角的
equilibration 英 [ˌiːkwɪlaɪˈbreɪʃən] 美 [ˌiːkwɪlɪˈbreɪʃən] *n.* 平衡；对消
coordinate 英 [kəʊˈɔːdɪneɪt] 美 [koˈɔrdəˌnet] *vt.* 使协调；使调和；整合
robust 英 [rəʊˈbʌst] 美 [roʊˈbʌst] *adj.* 精力充沛的

Useful Expressions

1. dismutase 英 [dɪsˈmjuːteɪs] 美 [dɪsˈmjuːteɪs] *n.* 岐化酶
The advantages and disadvantages of some measure methods for superoxide dismutase are compared.
对数种超氧化物歧化酶检测方法进行了比较，以分析其优缺点。

2. dilate 英 [daɪˈleɪt] 美 [daɪˈlet, ˈdaɪˌlet] *vi.* 膨胀；扩大；详述
At night, the pupils dilate to allow in more light.
到了晚上，瞳孔就会扩大以接收更多光线。

3. dock 英 [dɒk] 美 [dɑːk] *vt.* （使）船停靠码头；剪短（尾巴等）；削减；缩减
He threatens to dock her fee.
他威胁要扣掉她的服务费。

4. solvate 英 [sɒlˈveɪt] 美 [ˈsɔlvet] *vt.* 使……成溶剂化物
Ergometrine can be separated from the other ergot alkaloids with chloroform in which the former is insoluble as solvated crystals.

以后利用麦角新碱能与氯仿形成复合结晶,却又难溶于氯仿的特性,可以使麦角新碱与其他麦角生物碱分离。

5. configuration 英 [kənˌfɪɡəˈreɪʃn] 美 [kənˌfɪɡjəˈreɪʃn] n. 配置;[化](分子中原子的)组态;排列;[物]位形;组态

Prices range from \$119 to \$199, depending on the particular configuration.

价格因具体配置而异,从119美元至199美元不等。

Questions

1. How terrible are cardiovascular diseases?
2. What major challenges have studies on CDF's pharmacological effect confronted?
3. How was the compound-target associations related with CVD validated?

Chapter Ten

Introduction to Medicinal Plant Species

Passage 1

Preventive Effects of Dendrobium Candidum Wall ex Lindl. on the Formation of Lung Metastases in BALB/c Mice Injected with 26-M3.1 Colon Carcinoma Cells (Extract)

Results

In Vivo Anti-metastatic Effects of D. Candidum

26-M3.1 colon carcinoma cells were used to evaluate the anti-metastatic effects of D. candidum in vivo. Prophylactic inhibition of tumor metastasis by D. candidum was evaluated by using an experimental mouse metastasis model. D. candidum-treated mice had significantly fewer lung metastatic colonies as compared with control mice (number of metastatic tumors, 62 ± 6; $P<0.05$). A dose of 400 mg/kg b.w. D. candidum was the most effective at inhibiting lung metastasis. This concentration (inhibitory rate, 64.5%; number of metastatic tumors, 22 ± 3) inhibited tumor formation and lung metastasis to a greater degree as compared with a dose of 200 mg/kg solution (inhibitory rate, 46.8%; number of metastatic tumors, 33 ± 4).

IL-6, IL-12, TNF-α and IFN-γ Serum Levels in Mice

The control mice showed the highest serum levels of IL-6, IL-12, TNF-α and IFN-γ. These levels were significantly decreased in D. candidum-treated mice ($P<0.05$). A higher concentration of 400 mg/kg b.w. D. candidum was more effective as compared with the 200 mg/kg b.w. dose in promoting a decrease in cytokine serum levels.

Apoptosis-related Gene Expression of Bax and Bcl-2 in the Lung

To determine which apoptotic pathways were induced by D. candidum, the mice were treated with 200 and 400 mg/kg b.w. dose, and the lung tissues were dissected and analyzed for apoptosis-related gene expression by RT-PCR and western blotting. Evidence shows that in the presence of D. candidum, there were significant differences ($P<0.05$) in the expression of Bax and Bcl-2, with an increase in Bax expression and a reduction in Bcl-2 expression, as determined by RT-PCR. The Bax gene expression increased with D. candidium treatment, in a dose dependent manner, and Bcl-2 gene expression showed a crosscurrent when mice were treated with D. candidium ($P<0.05$). The results suggested that D. candidum induced apoptosis in 26-M3. 1 cell-injected lung metastatic mice through a Bax- and Bcl-2-dependent pathway. The increased expression of Bax and decreased expression of Bcl-2 induced by 400 mg/kg D. candidum was more notable at the mRNA expression level, as compared with the 200 mg/kg dose. From these results, D. candidum showed good anticancer effects in its ability to induce apoptosis, and these effects were observed in a dose-dependent manner.

Metastasis-Related Gene Expression of MMPs and TIMPs in the Lung

RT-PCR and western blot analysis was conducted to investigate whether the inhibitory effects of D. candidum on metastasis were due to gene regulation of metastatic mediators, such as MMPs (MMP-2 and MMP-9) and TIMPs (TIMP-1 and TIMP-2). The expression of MMPs and TIMPs was therefore analyzed in lung tissues taken from control and D. candidum-treated mice. Evidence shows that D. candidum significantly decreased the mRNA and protein expression of MMP-2 and MMP-9 ($P<0.05$), and significantly increased the expression of TIMP-1 and TIMP-2 ($P<0.05$). The most prominent anti-metastatic effects were associated with the most marked decrease in expression of MMP-2 and MMP-9, together with the most marked increase in expression of TIMP-1 and TIMP-2. These results suggested that the higher 400 mg/kg dose of D. candidum, cultivated in the presence of sulfur, could elicit a stronger anti-metastatic activity as compared with the lower 200 mg/kg dose.

Content of the D. Candidum Leaf

Eleven compounds were isolated and identified from the D. candidum leaf. Compound 1 was obtained as a clear crystal, and the 1H-NMR spectrum of this compound was as follows: δ 6.92 (2H, d), 6.62 (2H, d), 6.06 (2H, s), 6.03 (1H, s), 2.65 (4H, m). This compound was confirmed as dihydrogen resveratrol. Compound 2 was obtained as a white powder, and the 1H-NMR spectrum of this compound was as follows: δ 6.98 (2H, d), 6.74 (2H, d), 6.62 (1H, s), 6.47 (1H, d), 4.83 (1H, d), 4.63 (1H, d), 3.1—3.8 (12H), 3.73 (3H, s), 3.69 (3H, s), 2.74 (4H, m). This compound was confirmed as dendromoniliside E. Compound 3 was obtained as a black red needle, and the 1H-NMR spectrum of this compound was as follows: δ 11.00 (1H, s), 8.15 (1H, d), 6.06 (2H, s), 8.07 (1H, d), 6.95 (1H, s), 6.83 (1H, s), 6.15 (1H, s), 3.96 (3H, s), 3.93 (3H, s). This compound was confirmed as denbinobin. Compound 4 was obtained as a colorless needle, and the 1H-NMR spectrum of this compound was as follows: δ 4.72 (2H, m), 3.85 (1H, d), 6.06 (2H, s), 2.53 (1H, d), 2.49 (1H, t), 2.39 (1H, dd), 2.21 (1H, dd), 1.64 (1H, m), 1.35 (3H, s), 1.03 (3H, d), 0.95 (3H, d). This compound was confirmed as aduncin. Compound 5 was obtained as a white needle, and the 1H-NMR spectrum of this compound was as follows: δ 8.25 (1H, s), 8.10 (1H, s), 5.90 (1H, d), 4.66 (1H, dd), 3.5—4.2 (4H, m). This compound was confirmed as adenosine. Compound 6 was obtained as a white powder, and the 1H-NMR spectrum of this compound was as follows: δ 7.95 (1H, d), 5.85 (1H, d), 5.66 (1H, d), 3.2—4.3 (5H, m). This compound was confirmed as uridine. Compound 7 was obtained as a clear crystal, and the 1H-NMR spectrum of this compound was as follows: δ 10.60 (1H, s), 7.92 (1H, s), 6.45 (2H, s), 5.66 (1H, d), 3.4—4.4 (5H, m). This compound was confirmed as guanosine. Compound 8 was obtained as a white powder, and the 1H-NMR spectrum of this compound was as follows: δ 7.65 (1H, d), 7.41 (2H, d), 6.85 (2H, d), 6.33 (1H, d), 4.17 (2H, t), 1.69 (2H, m), 1.25 (54H, m), 0.85 (3H, t). This compound was confirmed as defuscin. Compound 9 was obtained as a white powder, and the 1H-NMR spectrum of this compound was as follows: δ 7.45 (2H, d), 6.82 (2H, d), 6.81 (1H, d), 5.83 (1H, d), 4.16 (2H, t), 1.67 (2H, m), 1.23 (54H, m), 0.88 (3H, t). This compound was confirmed as n-triacontyl cis-p-coumarate. Compound 10 was obtained as a white powder, and the 1H-NMR spectrum of this compound was as follows: δ 2.35 (2H, t), 1.62 (2H, m), 1.25 (24H, m), 0.88 (3H, t). This compound was confirmed as hexadecanoic acid. Compound 11 was obtained as a white powder, and the 1H-NMR spectrum of this compound was as follows: δ 3.85 (2H, t), 1.75 (2H, m), 1.45 (2H, m), 1.22 (54H, m), 0.85 (3H, t). This compound was confirmed as hentriacontane.

Discussion

Although D. candidum has been used as a traditional Chinese medicine, there has been little scientific research regarding its mechanism of action. D. candidum contains high concentrations of benzenes and their derivatives, phenolic, lignans, lactone, flavonoids and 18 novel D. candidum pigments. D. candidum has been previously reported to have various therapeutic effects on numerous pathological conditions, including inflammation, immunity, hyperglycemia and cancer.

Lower levels of IL-6, IL-12, IFN-γ and TNF-α cytokines are indicative of improved anticancer effects. IL-6 is regarded as an important tumor-promoting factor in various types of human cancer. An increased expression of IL-6 has been found in patients with cancer, in serum and tumor tissue. IL-12 has been shown to contribute to tumor eradication, through IFN-γ-dependent induction of the anti-angiogenic factors interferon-inducible protein 10 and monokine induced by gamma interferon. In addition, a previous study has shown that drugs targeting TNF-α may be useful for the treatment of cancers. In the present study, it was observed that the levels of IL-6, IL-12, TNF-α and IFN-γ in mice injected with 26-M3.1 cells were markedly decreased following D. candidum treatment. Based on this study, D. candidum showed a strong preventive effect on the development of lung metastases.

Tumor cells are able to migrate to another site, penetrate the vessel walls, continue to multiply and eventually form another tumor. Colon 26-M3.1 carcinoma cells have been previously used to evaluate anti-metastatic effects in vivo. Based on in vivo data from previous studies, 26-M3.1 colon carcinoma cells were used to examine the effects of D. candidum on metastasis in mice. The results further proved the activity of D. candidum, and the observed anticancer effects occurred in a dose-dependent manner.

Apoptosis is a fundamental cell event, and understanding its mechanisms of action will have a significant effect on antitumor therapy. The Bcl-2 family, which includes promoters (Bax and Bid) and inhibitors (Bcl-2 and Bcl-xL), is a key regulator in mitochondria-mediated apoptosis. In the present study, the gene and protein expression of Bax was increased, whereas the protein expression of Bcl-2 was decreased following treatment with D. candidum. Based on the gene expression results, D. candidum showed strong activity in promoting apoptosis in cancer.

MMPs comprise a family of zinc-dependent endopeptidases that function in tumorigenesis and metastasis. MMPs can cleave the majority of all extracellular matrix (ECM) substrates. Degradation of the ECM is a key event in tumor progression, invasion and metastasis. Among the MMP family members, MMP-2 and MMP-9 are molecules important for cancer invasion, and have been shown to be highly expressed in cancer cells. Inhibition of MMP activity is useful for controlling tumorigenesis and metastasis. TIMPs are naturally occurring inhibitors of MMPs that prevent catalytic activity by binding to activated MMPs, thereby blocking the degradation of the ECM. Disturbances in the ratio between MMPs and TIMPs have been observed during tumorigenesis. Maintaining the balance between MMPs and TIMPs, or increasing TIMP activity, are useful methods by which to control tumor metastasis. In the present study, strong anti-metastatic effects were correlated with a reduction of MMPs and an increase of TIMPs following administration of D. candidum in mice. From the results, D. candidum showed a strong anti-metastatic effect and, therefore, may be a functional drug for cancer prevention.

Numerous compounds were isolated and identified by NMR of the D. candidum leaf in the present study. Resveratrol is an antioxidant that has been often recommended for use as treatment in patients with colon cancer. Denbinobin is a biologically active chemical that has been demonstrated to inhibit colon cancer growth both in vitro and in vivo. Aduncin is a unique component that has only been found in Dendrobium. Aduncin may have anticancer effects, but its function requires further research. Adenosine serves as a physiological regulator and acts as a cardio-, neuro- and chemo-protector, and as an immunomodulator. Adenosine has been shown to exert anticancer effects at certain concentration. When administered in combination with chemotherapy, adenosine can enhance the chemotherapeutic index and acts as a chemoprotective agent. The availability of uridine can alter the sensitivity of tumor cells to antimetabolites. Adenine is incorporated into polynucleotides and the two compounds have been identified as important cancer-associated substances. Defuscin, n-triacontyl cis-p-coumarate, hexadecanoic acid and hentriacontane have also shown functional activities in human health. Taken together, these compounds all exhibit anticancer activities, and with the high content of functional compounds, this may explain why D. candidum showed a functional effect in cancer prevention. The synergy of these bioactive components may increase the anticancer effects of D. candidum.

In summary, the present study found that D. candidum has potent in vivo anti-metastatic activity, particularly in the inhibition of in vivo tumor metastasis. The analysis of cytokine levels, as well as mRNA and protein expression, has provided a mechanistic basis for these functional effects and a scientific basis for the development of D. candidum in cancer therapy.

https://www.spandidos-publications.com/ol/8/4/1879

Vocabulary

metastasis 英 [mə'tæstəsɪs] 美 [mə'tæstəsɪs] n. 转移
cytokine serum   细胞因子血清
apoptosis 英 [ˌæpə'təʊsɪs] 美 [ˌæpə'toʊsɪs] n. (细胞)凋亡
apoptotic 英 [æpɒp'tɒtɪk] 美 [æpɒp'tɒtɪk] adj. [医] 细胞凋亡的
mediator 英 ['miːdieɪtə(r)] 美 ['midiˌetə] n. 调解人；调停者；传递者；中介物
dihydrogen resveratrol   二氢白藜芦醇
hentriacontane 英 ['henˌtraɪə'kɒnˌteɪn] 美 ['henˌtraɪə'kɒnˌteɪn] n. 三十一(碳)烷
benzene 英 ['benziːn] 美 ['bɛnˌzin, bɛn'zin] n. 苯
eradication 英 [ɪˌrædɪ'keɪʃn] 美 [ɪˌrædɪ'keɪʃn] n. 摧毁；根除
carcinoma 英 [ˌkɑːsɪ'nəʊmə] 美 [ˌkɑːrsɪ'noʊmə] n. 癌
synergy 英 ['sɪnədʒi] 美 ['sɪnərdʒi] n. 协同；配合
uridine 英 ['jʊərɪdiːn] 美 ['jʊrɪˌdiːn] n. 尿(嘧啶核)苷

Useful Expressions

1. cleave 英 [kliːv] 美 [kliv] v. 劈开；紧贴；迅速穿过；信守
They just cleave the stone along the cracks.
他们就是顺着裂缝把石头劈开。

Questions

1. What effects does D. candidum have?
2. How many components does D. candidum leaf have?
3. What effect does uridine possess?

Passage 2

Scutellaria

Description

Most are annual or perennial herbaceous plants from 5 to 100 cm tall, but a few are subshrubs; some are aquatic. They have four-angled stems and opposite leaves. The flowers have upper and lower lips. The genus is most easily recognized by the typical shield on the calyx that has also prompted its common name.

Traditional use

Skullcaps are common herbal remedies in systems of traditional medicine. In traditional Chinese medicine they are utilized to "clear away the heat-evil and expel superficial evils". Scutellaria baicalensis in particular is a common component of many preparations. Its root, known as Radix Scutellariae, is the source of the Chinese medicine Huang Qin. It has been in use for over 2,000 years as a remedy for such conditions as hepatitis, diarrhea, and inflammation. It is still in demand today, and marketed in volumes that have led to the overexploitation of the wild plant. Its rarity has led to an increase in price, and encouraged the adulteration of the product with other species of Scutellaria.

In North America, Scutellaria lateriflora was used in Native American medicine to treat gynaecological conditions. It became a common treatment in America for rabies. Today it is still a popular medicinal herb. It is widely available as a commercial product used in western herbalism to treat anxiety and muscle tension. The plant reportedly commands prices of $16 to $64 per pound dry weight.

Constituents and pharmacology

The main compounds responsible for the biological activity of skullcap are flavonoids. Baicalein, one of the important Scutellaria flavonoids, was shown to have cardiovascular effects in in vitro. Research also shows that Scutellaria root modulates inflammatory activity in vitro to inhibit nitric oxide (NO), cytokine, chemokine and growth factor production in macrophages. Isolated chemical compounds including wogonin, wogonoside, and 3,5,7,2',6'-pentahydroxyl flavanone found in Scutellaria have been shown to inhibit histamine and leukotriene release. Other active constituents include baicalin, apigenin, oroxylin A, scutellarein, and skullcapflavone.

Some Scutellaria species, including S. baicalensis and S. lateriflora, have demonstrated anxiolytic activity in both animals and humans. A variety of flavonoids in Scutellaria species have been found to bind to the benzodiazepine site and/or a non-benzodiazepine site of the GABAA receptor, including baicalin, baicalein, wogonin, apigenin, oroxylin A, scutellarein, and skullcapflavone II. Baicalin and baicalein, wogonin, and apigenin have been confirmed to act as positive allosteric modulators and produce anxiolytic effects in animals, whereas oroxylin A acts as a negative allosteric modulator (and also, notably, as a dopamine reuptake inhibitor). As such, these compounds and actions, save oroxylin A, are likely to underlie the anxiolytic effects of Scutellaria species.

https://en.wikipedia.org/wiki/Scutellaria

Vocabulary

scutellaria 英 [skjuːtəˈlærɪə] 美 [skjuːtəˈlærɪr] *n.* 黄芩（属）；美黄岑

aquatic 英 [əˈkwætɪk] 美 [əˈkwætɪk] *n.* 水生动植物；水上运动

skullcap 英 [ˈskʌlkæp] 美 [ˈskʌlˌkæp] *n.* 美黄岑

scutellaria baicalensis　［医］黄芩

radix Scutellariae　＜拉＞黄芩

scutellaria lateriflora　侧花黄芩

gynaecological 英 [ˌɡaɪnəkəˈlɒdʒɪkl] 美 [ˌɡaɪnəkəˈlɑːdʒɪkl] *adj.* 妇科学的妇科的

baicalein 英 [ˈbeɪkælɪn] 美 [ˈbeɪkælɪn] *n.* 黄芩素

nitric oxide 英 [ˈnaɪtrɪk ˈɔksaɪd] 美 [ˈnaɪtrɪk ˈɑkˌsaɪd] 一氧化氮

benzodiazepine 英 [ˌbenzɒdaˈæzəˌpɪn] 美 [ˌbenzɒdaˈæzəˌpɪn] *n.* 苯（并）二氮（用于制造各种镇静药）

Useful Expressions

1. modulate 英 [ˈmɒdjuleɪt] 美 [ˈmɑːdʒəleɪt] *v.* 调节；调制；调整；变调
These chemicals modulate the effect of potassium.
这些化学物质可以调节钾的功效。

2. inflammatory 英 [ɪnˈflæmətri] 美 [ɪnˈflæmətɔːri] *adj.* 令人激动的；炎性的；发炎的
She described his remarks as irresponsible, inflammatory and outrageous.
她称他的话不负责任、具煽动性且极端无礼。

Questions

1. What the traditional use of scutellaria?
2. What was scutellaria used for in North America?
3. How many species does scutellaria have?